DEFINITIONS AND FORMULAE
IN PHYSICS

Definitions and Formulae in Physics

AMAN RAO

ANMOL PUBLICATIONS PVT LTD
New Delhi-110 002

ANMOL PUBLICATIONS PVT LTD
4374/4B, Ansari Road, Daryaganj
New Delhi-110 002.

Definitions and Formulae in Physics
Copyright © Reserved with Publisher
First Edition 1996
ISBN 81-7488-159-X

PRINTED IN INDIA

Published by J. L. Kumar for Anmol Publications Pvt Ltd New Delhi and Printed at Mehra Offset Press, Delhi.

Preface

This unique book entitled *Definitions and Formulae in Physics* is an essential reference book for all students in pursuit of examination success. It is meant as a capsule for last-minute revision before the examination. It is an attempt to give all possible definitions and formulae, and their applications in physics to acquaint students appearing in school and college examinations. It covers all the main branches of physics, including units, mechanics, heat and thermodynamics, sound, optics, nuclear physics, electricity and magnetism, etc. Thorough cross-referencing enhances definitions where further information is available. This book is marked by authenticity. It disseminates relevant, topical and latest information. This book will be welcomed by readers who are generally at a loss in remembering the common formulae near the examination.

Aman Rao

Contents

	Preface	v
1.	Measurement, Units and Dimensions	1
2.	Motion of a Body Along a Straight Line	14
3.	Newton's Laws of Motion	20
4.	Work, Power and Energy	29
5.	Friction	35
6.	Gravitation, Acceleration Due to Gravity and Satellites	41
7.	Uniform Circular Motion	51
8.	Rotational Motion	59
9.	Elasticity	64
10.	Fluid Motion and Pressure in a Fluid	67
11.	Surface Tension and Viscosity	74
12.	Thermometry and Thermal Expansion	80
13.	Kinetic Theory of Gases	89
14.	Calorimetry and Change of State	95
15.	Thermodynamics	99
16.	Isothermal and Adiabatic Changes	108
17.	Transmission of Heat and Thermal Conduction	110
18.	Simple Harmonic Motion	115

19.	Wave Motion and Speed of Mechanical Waves	121
20.	Vibrations of Strings and Air Columns	134
21.	Musical Sound and Doppler Effect	139
22.	Reflection and Refraction of Light	144
23.	Image Formation by Curved Mirrors and Lenses	150
24.	Defects of Eye and Optical Instruments	157
25.	Dispersion of Light and Chromatic Aberration	160
26.	Optical Instruments and Miscellaneous Concepts on Light	167
27.	Magnetism	181
28.	Properties of Magnetic Materials	186
29.	Electric Charge, Electric Field and Potential, Coulomb's Law	190
30.	Ohm's Law and Kirchoff's Law	200
31.	Heating Effect of Current	209
32.	Magnetic Effect of Current	216
33.	Galvanometer and Measuring Instruments	224
34.	Faraday's Law of Electrolysis	228
35.	Electromagnetic Induction	230
36.	Cathode Rays and Bohr's Model of Atom	235
37.	Diodes and Triodes	242
38.	X-Rays and Photoelectric Effect	253
39.	Radioactivity	258
40.	Nucleus and Nuclear Energy	262
41.	Universe	267
42.	Solids	277

1

Measurements, Units and Dimensions

1. Physics
It is the branch of science that deals with the fundamental aspects of matter and energy and these are convertible.

2. Fundamental Quantities and Units
The physical quantities which do not depend on other quantities are called **fundamental quantities**. The internationally employed fundamental quantities are those of *mass, length* and *time*. The units of such fundamental quantities are called **fundamental units**. The fundamental units are six independent units namely *length, mass, time, ampere, temperature* and *luminous intensity*.

3. Derived Units
The physical quantities which do not depend upon other quantities are known as *fundamental quantities*. The internationally employed *fundamental quantities* are those of *mass, length* and *time*. The units of the physical quantities which depend upon fundamental units or they can be derived from

fundamental units are known as *derived units* and quantities are known as *derived quantities*.

4. Various Systems of Units

Following systems of units are commonly used :

(a) C.G.S. (Centimetre-Gramme-Second) system.

(b) F.P.S. (Foot-Pound-Second) system.

(c) M.K.S. (Metre-Kilogramme-Second) system.

(d) S.I. (International System).

(a) **C.G.S. System :** In this system, the unit of length is centimetre, the unit of mass is gramme and that of the time is the second.

(b) **F.P.S. System :** In this system, the units of the length, mass and time are foot, pound and second respectively.

(c) **M.K.S. System :** In this system, the units of length, mass and time are metre, kilogramme and second respectively.

1 metre is the distance that contains 1650763.73 wavelength of orange-red light of Kr-86.

1 kilogram is the mass of 5.0188×10^{25} atoms of carbon-12.

1 second is the time in which cesium atom vibrates 91.92631770 times in an atomic clock.

(d) **S.I. System :** In this system, there are seven fundamental quantities.

5. Symbols of Some Quantities

Quantity	Symbol
Length	l, s
Mass	m
Time	t
Momentum	p
Moment of Inertia	I, J
Angular Momentum	L
Force	F
Energy of Work	E, W
Power	P
Pressure of Stress	p
Surface Tension	T, σ
Viscosity	n
Frequency	v, f
Torque	τ
Coefficient of Elasticity	q or Y, K, n

The three fundamental quantities chosen in Mechanics, namely mass, length and time are represented by [M], [L] and [T] respectively. How a certain derived quantity depends on the fundamental quantities can be expressed in terms of these symbols. For example from definition,

$$\text{Speed} = \frac{\text{Change in length}}{\text{Change in time}} = \frac{[L]}{[T]}$$

$$= [LT^{-1}]$$

The term dimension is used to denote this dependence on the basic quantities. Thus the dimension of speed is

$[LT^{-1}]$. The dimensions and units of various physical quantities are given in the table below:

Dimensions	SI Units
L	metre (m)
M	kilogramme (kg)
T	second (s)
MLT^{-1}	$kg\ ms^{-1}$
ML^2	$kg\ m^2$
ML^2T^{-1}	$kg\ m^2\ s^{-1}$
MLT^{-2}	newton (N)
ML^2T^{-2}	joule (J)
ML^2T^{-3}	watt (W)
$ML^{-1}T^{-2}$	pascal (Pa)
MT^{-2}	$N\ m^{-1}$ or Jm
$ML^{-1}T^{-1}$	$kg\ m^{-1}\ s^{-1}$
T^{-1}	hertz (Hz)
ML^2T^2	Nm
$ML^{-1}T^2$	Nm^{-2}

6. Fundamental Quantities in S.I. System

S.N.	Physical quantity	Name	Symbol
1	Length	metre	m
2	Mass	kilogramme	kg
3	Time	second	s
4	Temperature	kelvin	k
5	Luminous Intensity	candela	cd
6	Electric Current	ampere	A
7	Amount of substance	mole	$mol.$

There are two supplementary units in this system. They are

1. Plane angle   radian   rad.
2. Solid angle   steradian   Sr.

7. Characteristics of Physical Standard

Source of the characteristics of a physical standard are:—

1. Invariability,   2. Indestructibility
3. Accessibility,   4. Reproductivity

8. Abbreviation for Multiples and Sub-Multiples

The following table shows the abbreviation for multiples and sub-multiples :

Symbol	Prefix	Multiplier
T	tera	10^{12}
G	giga	10^{9}
M	mega	10^{6}
k	kilo	10^{3}
h	hecto	10^{2}
da	deca	10^{1}
d	deci	10^{-1}
c	centi	10^{-2}
m	milli	10^{-3}
μ	micro	10^{-6}
n	nano	10^{-9}
p	pico	10^{-12}

The following quantities have no dimensions : Specific gravity, Poisson's ratio, strain and angle.

Physical quantities can be classified as follows :

(i) *Dimensional Variable* : These are quantities like acceleration, velocity, force, etc.

(ii) *Dimensional Constants* : Quantities which have a constant value and yet have dimensions, e.g. the Gravitational Constant G; the velocity of light in vaccum, c : Planck's Constant, h.

(iii) *Non-dimensional Variables* : These are variables but have no dimensions, e.g. strain, angle. Quantities which have zero dimension in each of the fundamental quantities are called *numerics*.

Quantity	Symbol	Dimension	SI Units
Area	A, S	L^2	m^2
Volume	V	L^3	m^3
Density	ρ	ML^{-3}	$kg\ m^{-3}$
Velocity	u, v	LT^{-1}	$m\ s^{-1}$
Acceleration	a	LT^{-2}	$m\ s^{-2}$

(iv) *Non-dimensional Constants* : These are mere numbers like 5, 7, π, etc.

9. Principle of Homogeneity of Dimensions

According to this principle, only dimensionally alike quantities can be added or subtracted. In other words, if there is an equation containing a number of terms to be added or subtrated, all terms to be added or subtracted should have the same dimensions e.g. in the equation

$$S = ut + \frac{1}{2}at^2$$

the three terms S, ut and $\frac{1}{2}at^2$ all have the same dimensions.

Definitions and Formulae in Physics

10. Uses of the Principle

(i) *Checking of results*—If there be any given results in the form of an equation, one can verify their correctness by finding out dimensions of different terms to be added or subtracted.

(ii) *Finding out conversion factors*—knowing the basic ratios of the units of mass, length and time in any two systems, one can find out conversion factors of the units of any quantity in these systems, by the use of dimensional considerations.

(iii) *Derivation of results*—By dimensional considerations one can derive simple power relations amongst quantities. However complicated relations involving dependence on more than three quantities or involving. T-ratios, log, exponential etc., cannot be derived.

The time period of a simple pendulum may possibly depend upon the following factors :·

(i) mass of bob (m), (ii) length of simple pendulum (l), (iii) acceleration due to gravity (g) and (iv) angle of swing of simple pendulum (θ).

Let
$$t = K m^a \cdot l^b \cdot g^c \cdot \theta^d$$

Where K is constant of proportionality.

Taking dimension on both sides, we have

$$[T] = [M]^a [L]^b [LT^{-2}]^c$$

or
$$T = M^a L^{b+c} \cdot T^{-2c}$$

Equating powers of M, L and T on both sides $a = 0$, $b + c = 0$ and $-2c = 1$

$$\therefore \quad c = -\frac{1}{2}, b = \frac{1}{2} \text{ and } a = 0$$

Thus $t = K\, l^{1/2} g^{-1/2}$

$$t = K\sqrt{\left(\frac{l}{g}\right)}$$

It is observed experimentally that $K = 2\pi$

$$\therefore \quad t = 2\pi \sqrt{\left(\frac{l}{g}\right)}$$

11. Limitation of Dimensional Analysis

Some of the limitations of dimensional anslysis are :—

(1) It provides no information about pure numerics.

(2) The dimensional analysis cannot be used to derive a relation if the physical quantity depends upon more than three factors because we can get only three equations by equating the powers of M, L and T.

(3) It is restricted to relationships that are power functions.

The relationship involving trigonometrical functions or exponentials cannot be derived.

12. Percentage Error

If Δx is the error in the measurement x, then fractional error = $\Delta x/x$

and percentage error = $(\Delta x/x) \times 100$

Experimental percentage error =

$$\frac{\text{Experimental value} \approx \text{Standard value}}{\text{Standard value}} \times 100$$

Definitions and Formulae in Physics

13. Significant Figures

The significant figure is defined as the figure in a given number which can be realised. For example, if the radius measured by screw gauge is 3.423 cm, the number of significant figures in the measurement is 4. Greater is the number of significant figures in a measurement, smaller is the percentage error. Here it should be remembered that the limit of accuracy of a measuring instrument is equal to the least count of the instrument.

14. Percentage Error

If Δx is the error in a measurement x, then fractional error

$$= \frac{\Delta x}{x}$$

and %age error $= \dfrac{\Delta x}{x} \times 100$

Experimental %age error =

$$\frac{\text{(Experimental value)} - \text{(Standard value)}}{\text{Standard value}} \times 100$$

Maximum %age error

in $y = a \times b$ is given by

$$\left[\frac{\Delta y}{y}\right]_{\max} \times 100 = \frac{\Delta a}{a} \times 100 + \frac{\Delta b}{b} \times 100$$

15. Scalar Quantity

It is a quantity which has magnitude but no direction, such as volume, time, mass, etc.

16. Vector Quantity

It is a quantity which has both magnitude and direction, such as velocity.

17. Representation of a Vector

A straight line such as OA with an arrow is used to represent a vector.

The direction of the line indicates the O → A direction and its length represents the magnitude of the given vector. The sense in which a given vector acts (from O to A) is specified by the arrow.

18. Unit Vector

The vector of unit magnitude is known as unit vector. The unit vector in the direction of vector $\vec{A}$ is defined as follows

$$\hat{a} = \frac{\vec{A}}{|\vec{A}|}$$ where $|\vec{A}|$ is known as mode and represents the magnitude of vector $\vec{A}$.

19. Orthogonal Unit Vectors

The unit vector along three mutually perpendicular directions i.e., x, y and z axes are called orthogonal unit vectors. These are represented by $\hat{i}, \hat{j},$ and $\hat{k}$ respectively.

20. Resolution of a Vector in two dimensions

If a vectors $\vec{A}$ makes an angle θ with horizontal x-direction and its components along two mutually perpendicular directions (i.e. along x and y axis) are $\vec{A} = \vec{A_x} + \vec{A_y}$ and in terms of unit vectors

$$\vec{A} = A_x \cdot \hat{i} + A_y \cdot \hat{j},$$

where A_x = Magnitude of Horizontal component $\vec{A_x} = A \cos θ$; Magnitude of vertical component $\vec{A_y} = A \sin θ$.

$$A = \sqrt{A_x^2 + A_y^2} \text{ and } θ = \tan^{-1}(A_y/A_x)$$

21. Resolution of a Vector in three Dimensions

If the components of vector $\vec{A}$ along three mutually perpendicular directions x, y and z axis are $\vec{A_x}$, $\vec{A_y}$ and $\vec{A_x}$, then

$$\vec{A} = \vec{A}_x + \vec{A}_y + \vec{A}_z$$

or
$$\vec{A} = A_x \cdot \hat{i} + A_y \cdot \hat{j} + A_z \cdot \hat{k}.$$

and
$$A = \sqrt{A_x^2 + A_y^2 + A_z^2}.$$

N.B. If $\vec{A_x}$, $\vec{A_y}$ and $\vec{A_z}$ make the angle α, β and γ with x, y and z axis respectively, then $A_x = A \cos \alpha$, $A_y = A \cos \beta$, $A_z = A \cos \gamma$ and are called direction cosines of the vectors.

22. Scalar Product or Dot Product of two Vectors

It refers to the product of the magnitudes of two vectors and the cosine of the angle between them. Thus, if $\vec{A}$ and $\vec{B}$ are two vectors, their scalar product written as $\vec{A} \cdot \vec{B}$ is given by

$$\vec{A} \cdot \vec{B} = AB \cos \theta$$

where A is the magnitude of $\vec{A}$, B the magnitude of $\vec{B}$ and θ is the angle between two vectors meausred from $\vec{A}$ to $\vec{B}$.

Since A and B are scalars and $\cos \theta$ is a pure number, *the dot product of two vectors is a scalar quantity.*

Example. Work, $W = \vec{F} \cdot \vec{d}$ where F = force and d = displacement.

23. Vector Product of two Vectors

It is defined as the vector normal to the plane including the given (two) vectors and having a magnitude equal to the product of the magnitudes of given vectors and sine of the angle between them. Thus, if $\vec{A}$ and $\vec{B}$ are two vectors, their vector product written as $\vec{A} \times \vec{B}$ is given by

$$\vec{C} = \vec{A} \times \vec{B} = AB \sin \theta$$

where A is the magnitude of $\vec{A}$, B is the magnitude of $\vec{B}$ and θ is the angle between the two vectors measured from $\vec{A}$ to $\vec{B}$. The direction of C is perpendicular to the plane, containing A and B and is given by right hand screw rule.

Example (i) **Torque** is the vector product of position vector and force vector

$$\vec{\tau} = \vec{r} \times \vec{F}$$

(ii) **Angular momentum** is the vector product of position and linear momentum vector

$$\vec{L} = \vec{r} \times \vec{p}$$

(iii) **Linear Velocity** is the vector product of angular velocity vector and position vector

$$\vec{v} = \vec{w} \times \vec{r}$$

24. Scalar Product of Unit Vectors :

$$\hat{i} \cdot \hat{i} = 1, \hat{j} \cdot \hat{j} = 1, \hat{k} \cdot \hat{k} = 1$$

$$\hat{i} \cdot \hat{j} = 0, \hat{j} \cdot \hat{k} = 0, \hat{k} \cdot \hat{i} = 0$$

25. Vector Product of Unit Vectors :

$$\hat{i} \times \hat{i} = \hat{j} \times \hat{j} = \hat{k} \times \hat{k} = 0$$

$$\hat{i} \times \hat{j} = 1, \hat{j} \times \hat{k} = 1 \text{ and } \hat{k} \times \hat{i} = 1$$

$$\hat{j} \times \hat{i} = -1, \hat{k} \times \hat{j} = -1 \text{ and } \hat{i} \times \hat{k} = 1$$

If
$$\vec{A} = A_x \times \hat{i} + A_y \times \hat{j} + A_z \times \hat{k}$$

$$\vec{B} = B_x \times \hat{i} + B_y \times \hat{j} + B_z \times \hat{k}$$

Then,
$$\vec{A} \cdot \vec{B} = A_x \times B_x + A_y \times B_y + A_z \times B_z$$

$$= \hat{i}(A_y \cdot B_z - B_y \cdot A_z) - \hat{j}(A_z \cdot B_x - B_z \cdot A_x)$$
$$+ \hat{k}(A_x \cdot B_y - B_x \cdot A_y)$$

2
Motion of a Body Along a Straight Line

1. Motion

Acceleration is defined as the rate of change of velocity of a body. The acceleration of the body is said to be *uniform acceleration* when its velocity changes by equal amounts in equal invervals of time.

(i) *Motion in a straight line.* In this type of motion, the direction of the particle's velocity remains un-changed but the magnitude changes uniformly. If the magnitude of particle's velocity increases with time then the motion is called *accelerated motion*. If the magnitude of particle's velocity decreases with time the acceleration is negative and the motion is called *retarded motion*.

(ii) *Projectile motion.* The acceleration has same size and direction i.e., acceleration due to gravity.

(iii) *Particle moving in a circle with constant speed.* In this type of motion; the acceleration is constant in magnitude but the direction is changing at a constant rate.

2. Distance and Displacement

The total length of the actual path covered by a body is known as *distance* traversed by the body. It is a scalar quantity and its unit in M.K.S. system is metre. The directional distance between initial and final positions of the body is known as *displacement*. It is a vector quantity and its unit in M.K.S. system is also metre. It should be remembered that the displacement is independent of actual path followed by the body. To differentiate between distance and displacement we consider the example of a body projected upward to a height h and returns to the point of projection. In this problem the total distance traversed by the body is $2h$ while the displacement of the body is zero.

3. Speed and Velocity

The distance covered by a body in unit time without any reference to the direction of motion is known as *speed*. It is scalar quantity and its unit is metre/sec. The distance travelled by a body in unit time in a specified direction is known as *velocity*. It is a vector quantity and its unit is also metre/sec.

4. Velocity, Average Velocity and Instantaneous Velocity

The *velocity* of a particle is the rate at which its position changes with time.

The *average velocity* for a particle during a time interval Δt is defined by

$$\bar{v} = \frac{\Delta r}{\Delta t} \quad (\bar{v} = \text{Average Velocity})$$

where Δr is displacement (a vector).

The velocity of a particle at any given instant of time is known as the instaneous velocity. The magnitude v of the instantaneous velocity is known as the speed and is simply absolute value of v.

$$\text{Instantaneous velocity} = \lim_{\Delta t \to 0} \frac{\Delta r}{\Delta t} = \frac{dr}{dt}$$

5. Acceleration

The rate of change of velocity of a body is known as acceleration. Its unit is metre/sec² and is a vector quantity.

$$\text{Average acceleration } a = \frac{\Delta v}{\Delta t}$$

and Instantaneous acceleration

$$\lim_{\Delta t \to 0} \frac{\Delta v}{\Delta t} = \frac{dv}{dt}$$

6. Displacement-time Graph

When the displacement of a body is plotted on Y-axis and time on the X-axis then the graph so obtained is called as displacement-time graph.

Characteristics of displacement time graph :

(i) If the graph is a straight line parallel to X-axis [see Fig. (1)], shown by line *ab,* it means that the body is at rest i.e., the velocity of the body is zero (ii) A straight line inclined to X-axis (such as *oc* and *fg*) shows that the body is moving with a constant velocity. It should be remembered that a straight line inclined to X-axis by an angle greater than 90° (line *fg*) represent negative velocity. (iii) It is important to note that no line can ever be perpendicular to the time axis because it represents infinite velocity, (iv) if the curve is of the type *od* whose slope decreases with time, the velocity goes on decreasing i.e., the motion is retarded. (v) If the

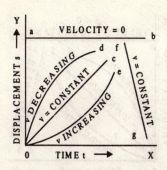

Fig. 1.

curve is of the type *oe* whose slope increases with time, the velocity goes on increasing i.e., the motion is accelerated.

7. Velocity-time Graph

When the time is plotted on X-axis and velocity on Y-axis, the graph so obtained is called as velocity-time graph.

Characteristics of velocity-time graph :

(i) If the graph is a straight line parallel to X-axis [see Fig. (2)] shown by line *ab,* it means that the body is moving with constant velocity or acceleration (a) is zero. (ii) If the graph is a straight line inclined to X-axis with positive slope (line *oc*), it reveals that the body is moving with constant acceleration. (iii) If the graph is a straight line inclined to X-axis with negative slope (line *fg*), it means that the body is under retardation, (iv) No velocity-time graph can be ever normal to the time axis because it will represent infinite acceleration. (v) If the graph obtained is a curve like *od* whose slope decreases with time, the acceleration goes on decreasing. (vi) If the graph obtained is a curve like *oe* whose slope increases with time, the acceleration goes on increasing. (vii)

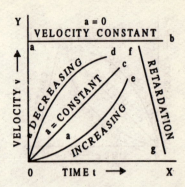

Fig. 2.

The area of velocity-time graph with time axis represents the displacement of that body.

8. Time-Acceleration Graph

If the acceleration of a body is plotted on Y-axis and time on X-axis, then the curve so obtained is called as acceleration-time graph.

Characteristic of acceleration time graph :

(i) When the graph is a parallel line to X-axis, then acceleration is constant.

(ii) When the graph is an oblique straight line having positive slope, then the acceleration is uniformly increasing.

(iii) When the graph is an oblique straight line having negative slope, then the acceleration is uniformly decreasing.

9. Equations of Motion

The equations of motion are :—

(1) $v = u + at$

(2) $s = ut + (1/2) at^2$

(3) $v^2 = u^2 + 2as$

where

u = initial velocity of the body.

v = final velocity of the body after time t.

s = distance travelled in time t.

a = uniform acceleration of body.

The distance covered by the body in n^{th} second

$$s_n = u + \frac{a}{2}(2n - 1).$$

3

Newton's Laws of Motion

1. Newton's Laws of Motion

Newton's laws of motion from the basis of Mechanics :

I. Every body persists in its state of rest or of uniform motion in a straight line unless it is compelled to change that state by forces imposed upon it.

II. Rate of change of momentum is proportional to the imposed force, and takes place in the direction of the straight line in which the force acts.

$\vec{F} = m\vec{a}$, where $\vec{F}$ is the (vector) sum of all the forces acting on a body, m is the mass of the body $\vec{a}$ is its (vector) acceleration.

If the forces and acceleration are resolved in three mutually perpendicular directions, the x, y, z axes.

$F_x = ma_x$
$F_y = ma_y$ and $F_z = ma_z$

III. To every action there is always opposite and equal reaction; or, the mutual reactions of two bodies upon each other are always equal, and directed to contrary parts.

The action and reaction forces which always occur in pairs *act on different bodies.* A body *X* exerts a force on body *Y*, and body *Y* oppositely directed force on body *X*.

It is possible to show that first law and third law can be derived from second law. Thus second law is the master law.

2. Importance/Application of these Laws

First law, discovered by Galileo initially, is called the law of Inertia. Inertia is a fundamental property of matter and according to it a body has always to be in the state in which it is kept initially unless acted upon by an external force.

Second law as already explained is the master law and calculations involving motion under a force or a number of forces are always to be based upon this law. Also, we get the concept and definition of force from this law only.

Thirdly law forms the principle of all propulsion motions e.g. motion of jet plane. This law is derivable from the principle of conservation of linear momentum.

3. Centre of Mass

If a system consists of a number of particles of masses m_1, m_2, m_3.... it is possible to simplify the application of Newton's law, to the system, by defining a point called the Centre of mass of the system.

Thus let the mass m, have the coordinates x_i, y_i, z_i with reference to a coordinate system. The coordinates of the centre of mass are :

$$X = \frac{\Sigma m_i x_i}{\Sigma m_i} = \frac{\Sigma m_i x_i}{M}$$

$$Y = \frac{\Sigma m_i y_i}{\Sigma m_i} = \frac{\Sigma m_i y_i}{M}$$

$$Z = \frac{\Sigma m_i z_i}{\Sigma m_i} = \frac{\Sigma m_i z_i}{M}$$

where $M = \Sigma m_i$, the total mass of the particles.

The centre of mass of system of particles depends only on the masses of the particles and the positions of the particles relative to one another.

A rigid body such as a metre-stick can be thought of as a system of closely-packed particles. Hence, it has also a centre of mass.

Under the action of external forces, the centre of mass of a system of particles moves as though the entire mass of the system were concentrated at the centre of mass and all the external forces were applied at the point. We can obtain the translational motion of a body, that is, the motion of the centre of mass, by assuming that the entire mass of body is concentrated at its centre of mass, and all the external forces are applied at the point.

4. Impulse

If a force F acts on a body for a duration Δt, then impulse is defined as the product of force and its time of action. Thus

Impulse = Force × duration = $F \times \Delta t$.

5. Weight of Body in a Lift

Earth attracts every body towards its centre. The force of attraction exerted by the earth on the body is known as force

Definitions and Formulae in Physics

of gravity. If m be the mass of the body then the gravitational force on it will be mg. The weight of a body is equal to the gravity force $W = mg$. But when the body is on an accelerated platform, the weight of the body appears changed. The new weight is known as apparent weight. The apparent weight of a man standing in a lift which is in motion is given by the following :—

(i) *When the lift is unaccelerated (i.e., $v = 0$ or constant).*

The situation is shown in Fig. (1a). In this case $R = ma = 0$.

Hence apparent weight

$$W' = \text{Actual weight} = mg \qquad ...(1)$$

(ii) *When the lift is accelerated upward.* In this case, there will be two forces acting on the man i.e., weight mg and reaction $R = ma$ both acting in the downward direction as shown in Fig. (1b).

∴ Apparent weight

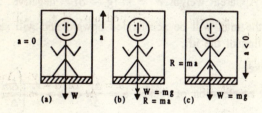

Fig. (1)

$$W' = mg + R + mg + ma = m(g + a) \qquad ...(2)$$

or Apparent weight W' > Actual weight W.

(iii) *When the lift is accelerated downward.* This situation is shown in Fig. (1c). Here the weight mg acts downward while the reaction $R = ma$ acts upward. We assume that $a < g$. Hence

$$\text{Apparent weight } W' = W - R = mg - ma$$
$$= m(g - a) \qquad \ldots(3)$$

∴ Apparent weight W' < Actual weight W.

Now we consider the special case when $g = a$.

In this case,

$$\text{Apparent weight } W' = 0$$

Thus in a freely falling lift, the man will experience a state of weightlessness.

(iv) *When the lift is accelerated downward such that $a > g$*

In this case $R = ma$ is greater than the weight mg.

∴ Apparent weight $W' = m(g - a)$ = negative

So the man will be accelerated upward and will stay at the ceiling of the lift.

6. Rocket

The acceleration of rocket, $a = \dfrac{\Delta v}{\Delta t}$ or $a = \dfrac{v_r}{m}\left(\dfrac{\Delta m}{\Delta T}\right) - g$, where $v_r\left(\dfrac{\Delta m}{\Delta t}\right)$ is known as *thrust of rocket*, v_r is the velocity of gas (ejected) related to rocket and Δm is the mass of gas ejected, m is the mass of the rocket and unburnt fuel. In outer space acceleration due to gravity becomes negligibly small or $g = 0$ and we have $a = \dfrac{v_2}{m}\left(\dfrac{\Delta m}{\Delta t}\right)$

(i) Rocket can go up in vacuum as reaction is not due to air but due to escaping vapours.

(ii) Acceleration increases as rocket moves upwards till the whole of the fuel is exhausted.

7. The Centre of Gravity of a Body

It is the point at which the entire weight of the body may be supposed to act.

8. Equilibrium of Body

A body is said to be in *translational equilibrium* if it is at rest or moving with uniform speed in a straight line. It is said to be in *rotational equilibrium*. If it is not rotating or if it is rotating at constant angular speed about an axis.

9. Conditions for the Equilibrium of a Rigid Body under the Action of Co-planar Forces

(1) The resultant of all forces in a given direction must be zero.

(2) The algebraic sum of moments of all forces acting about any axis perpendicular to the plane of the forces must be zero.

10. Equilibrium under three non-parallel Forces

Here the conditions are :

(1) The three forces must lie in a plane.

(2) Their lines of action must intersect in a common point.

(3) Vectors representing these forces in magnitude and direction when taken in order must form a closed triangle.

11. Unit of Force

In MKS system the unit of force is **newton**. It is the force which can impart an acceleration of 1 metre/sec^2 to a body of mass 1 kilogram. The unit is abbreviated as *nt*.

12. Apparent Weight

(i) Apparent weight when lift is moving up

If the lift is moving upward with an acceleration a then forces acting on man are his weight mg downward and reaction of lift on man, R, upward. R must be greater than mg since the lift is moving up. Hence, net force acting on man is $F = R - mg$ in upward direction but net force acting on man is $m.a.$ where a is the acceleration in the upward direction.

Hence, $$R - mg = m.a.$$

or $$R = mg + ma = m(g + a)$$

Therefore weight indicated by weighing machine will be equal to w',

$$w' = m(g + a)$$

Thus, when lift moves up the apparent weight increases.

Apparent weight when the lift is moving down

$$w' = m(g - a)$$

Thus apparent weight decreases when the lift is moving down.

13. Weightlessness

If the string of the lift (coming down) is broken then lift will start falling down like a freely moving body with acceleration g. Now apparent weight $w' = m(g - a)$

But $\qquad a = g$

Hence $\qquad w' = m(a - a) = 0$

Now pointer of weighing machine in lift will indicate zero weight. In other words, man will be weightless.

14. Momentum of a Body

It is the product of the mass of the body and its velocity. Its unit is kg m/s. It is vector having the same direction as the velocity.

15. Conservation of Momentum

In any collision between two or more bodies the vector sum of the momenta after impact is equal to that before impact.

16. Impulse

It is equal to the product of the force and the time for which it acts on the body. It is a vector having the same direction as the force. Its unit is newton-sec.

Impulse = change of momentum produced.

$Ft = (mv_2 - mv_1)$ where v_1 is the velocity before the force was applied and v_2 is the velocity after the force F has acted for time t.

Impulse = $(\vec{F} \times \Delta t)$ newton × sec.

17. Laws of Conservation of Linear Momentum

When the resultant external force acting on a system is zero, the total vector momentum of the system remains constant, or

$$(\vec{p_1} + \vec{p_2}) = \text{constant.}$$

or $\quad m_1 \vec{u_1} + m_2 \vec{u_2} = m_1 \vec{v_1} + m_2 \vec{v_2}$

where $\vec{u_1}$ and $\vec{u_2}$ are the velocities of the bodies of masses m_1, m_2 respectively before impact and $\vec{v_1}$, $\vec{v_2}$ are the respective velocities after impact.

4

Work, Power and Energy

1. Work

If the point of application of a force moves in the direction of the force. Work is said to be done. It is measured as $W = \vec{F} \cdot \vec{S}$, where $\vec{F}$ is the force and $\vec{S}$ is the distance moved, in the *S.I.* units of joules other with of work are ergs and electron both, we have,

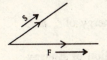

$$1 \text{ Joule} = 10^7 \text{ ergs}$$

and $1eV = 1.6 \times 10^{+19}$ joules

It is clear that if there is a motion under a central force e.g. circular motion under centripetal force, the distance moved and force applied being ⊥ar to each other, there would be no work done. Another example of motion under central force is planetary motion.

2. Power

It is the rate of doing work. It measured in watts. One watt is equal to 1 joule/sec. Other units are ergs/sec and kilowatts.

3. Energy

Energy is the capacity of a body to do work. It is measured in the same units as those of work. Mechanical Energy is of two types, viz., Potential and Kinetic Energies.

Potential Energy is the energy possessed by a body by virtue of its position e.g. a body kept at a height 'h' above the surface of the earth, when allowed to fall freely, can move done through a vertical distance 'h' under a vertical force of its own weight 'mg'. It is therefore, called to be having on energy mgh.

Kinetic Energy is the energy possessed by a body by virtue of its motion. It can be shown that if a body with mass 'm' and velocity 'v' be stopped to velocity 'zero', it will do an amount of work $\frac{1}{2}$ 'mv^2', before coming to rest. Thus, it possesses a kinetic energy of $\frac{1}{2} mv^2$.

4. Law of Conservation of Energy

For a system which involves no losses, total energy is conserved. Nevertheless, if may change its form. This means for a mechanical system, the sum total of potential and kinetic energies must be constant. This can be illustrated by taking an example of a freely falling body. Initially, when starting from rest, from a certain height, it will possess all potential energy and no kinetic energy. But as it proceeds, if starts loosing its potential energy on account of decreased height but its kinetic energy starts increasing on account of con-

Definitions and Formulae in Physics

tinuous increase in velocity. As any point in mid-path, it will partly have potential and partly kinetic energies. Later, when the body comes down, on the earth, it ceases to have any potential energy and all energy it possesses is kinetic. Throughout, the sum of potential and kinetic energies remains constant, although continuously potential energy is getting converted into kinetic one.

5. Other Forms of Energy

Besides mechanical energy, there one other forms of energy also e.g. electrical energy, magnetic energy, nuclear energy, chemical energy, light energy, heat energy, solar energy etc. These are all interconvertible.

6. Linear Momentum of a System of Particles

The total momentum P of a system of particles is the vector sum of the momentum of the individual particles. However, in virtue of the property of the centre of mass, the total momentum of a system of particles is equal to the product of the total mass of the system and the velocity of the centre of mass.

When the resultant external force acting on a system is zero, the total vector momentum of the system remains constant. This is called the principle of the conservation of linear momentum.

The momentum of individual particles may change but their sum remains constant if there is no net external force.

If a projectile following the usual parabolic trajectory, bursts during flight, the centre of mass of the fragments continues along the same parabolic path.

If two bodies of masses m_A and m_B move with velocities $\vec{u_A}$ and $\vec{u_B}$ before they collide, and have velocities $\vec{v_A}$ and $\vec{v_B}$ after collision.

$$\vec{mu_A} + \vec{mu_B} = \vec{mv_A} + \vec{mv_B}$$

If the velocities before and after collisions are in the same straight line,

$$mu_A + mu_B = mv_A + mv_B$$

However, if the collision results in motion in different directions, one can resolve the momentum in two mutually perpendicular directions and this law is applied separately in the two directions respectively.

For horizontal direction

$$m_1u_1 \cos \theta_1 + m_2u_2 \cos \theta_2 = m_1v_1 \cos \varphi_1 + m_2v_2 \cos \varphi_2$$

For vertical direction

$$m_1u_1 \sin \theta_1 + m_2u_2 \sin \theta_2 = m_1v_1 \sin \varphi_1 + m_2v_2 \sin \varphi_2$$

Whether there is conservation of kinetic energy or not, in the collision, conservation of linear momentum holds good.

In actual cases, there is always a certain loss of kinetic energy. According to an empirical result, between two given bodies, the velocity of separation bears a constant ratio to the velocity of approach. This constant is called as coefficient of restitution 'e'.

Thus, $$-e = \frac{v_A - v_B}{u_A - u_B}$$

If the collision is perfectly elastic, $e = 1$, and there is no loss of kinetic energy.

If the collision is *completely inelastic,* the bodies stick together on collision and move together, that is, there is no velocity of separation then and $e = 0$.

7. Einstein's Mass-energy Equivalence

According to Einstein neither mass nor energy of the universe is conserved but mass and energy are interconvertible. The conversion is expressed by Einstein's relation.

$$E = m c^2$$

where c is the velocity of light.

8. Newton's Experimental Law

When two bodies collide directly then the ratio of relative velocity after collision to the relative velocity before collision is a constant quantity. This constant quantity is expressed by the letter e. The relative velocity after collision is in opposite direction. When the two bodies collide obliquely, their relative velocity resolved along their common normal after collision bears a constant ratio to their relative velocity before collision resolved in the same direction and is of opposite sign. e is called the coefficient of restitution. Hence,

$$\frac{v_1 - v_2}{u_1 - u_2} = -e,$$

where u_1, u_2 are the velocities of two bodies before collision and v_1, v_2 are the velocities of the bodies after collision respectively. e lies between 0 and 1. When $e = 0$, the collision is said completely inelastic and when $e = 1$, the collision is said as perfectly elastic.

9. Elastic and Inelastic Collisions

(a) **Elastic collision.** *If the total kinetic energy of two bodies remain to be the same both after and before impact,* *the collision is said to be perfectly elastic.* Collisions between atomic, nuclear and fundamental particles are examples of elastic collision.

(b) **Perfectly inelastic collision.** The collision is known as perfectly inelastic collision when there is a loss of kinetic energy during collision and colliding bodies stick together and move as a single unit. For example, the collision between a bullet and a target is perfectly inelastic when the bullet remains embedded in the target. In this case kinetic energy is not conserved. Between the two limits of perfectly elastic and perfectly inelastic collisions, all other collisions are imperfectly elastic.

10. Loss of Kinetic Energy is Inelastic Collision

$$\frac{k_2}{k_1} = \frac{(m_1 + m_2) m_1^2 u_1^2}{m_1 u_1^2 (m_1 + m_2)^2} = \frac{m_1}{m_1 + m_2}$$

$\therefore m_1 < m_1 + m_2$, hence the R.H.S. is less than unity so the total kinetic energy decreases in inelastic collision.

5
Friction

1. Friction

If we consider the case of a body placed over a rough horizontal table and the body is pulled by a small horizontal force, it does not move. This shows that there is another horizontal force that opposes the applied pull. This opposing force is called the *friction force* and is exerted by table on the body. when the pulling force is increased, the body starts slipping. This is due to the fact that there is a limit to the magnitude of the frictional force. When the pulling force exceeds the maximum frictional force, the body accelerates according to Newton's law.

The property by virtue of which a resisting force is created between two rough bodies that resists the sliding of one body over the other is known as friction. The force that always acts in the direction opposite to that in which the body has a tendency to slide or move is known as force of friction.

The maximum frictional force between two surfaces is found to depend on nature of surface and normal contact force between two surfaces. It is independent of the area of contact.

The frictional force between the two surfaces before the relative motion actually starts is called *static friction*. The frictional force when the surfaces in contact are in relative motion is known as *kinetic (dynamics) friction*.

2. Frictional Force

Whenever two surfaces in contract have a relative motion between them, a force called frictional force comes into play, trying to oppose this relative motion. Frictional force chiefly depends upon the roughness of the two surfaces in contract. Limiting force of friction i.e., the maximum force of friction between two surface does not depend upon the area of contract. It is directly proportional to the normal reaction, if one takes the case of slipping a body on a given surface as shown. And as already mentioned it has to act in a direction so as to oppose the relative motion between the two surfaces.

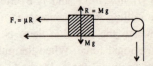

∴ Force of friction

$$F_f \propto R$$

⇒ $F_f = \mu R$. Here μ is a constant called the coefficient of friction between the two surfaces. For a horizontal motion $R = Mg$, but for motion along an inclined plane $R = Mg \cos \theta$.

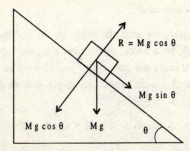

Force of friction can be of two types, viz., static and dynamic. Static friction, which comes into play when the motion is yet to start is more than dynamic friction which is effective when motion has already started. Further, dynamic friction is classified into sub-categories like slipping friction, rolling friction etc.

Friction is a necessary evil in the sense that it results in wastage of power applied but for transfer of motion from one body to another, as in a conveyor belt, it is necessary to have friction otherwise motion cannot be transferred.

3. Limiting Friction, Coefficient of Friction and Angle of Friction

Consider the case of a body which rests on a rough table as shown in Fig. (1). Its weight *mg* is acting downwards and

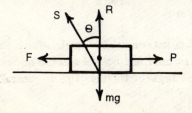

Fig. 1

normal reaction R is acting in opposite direction such that the two balance each other.

Now, suppose, we pull the body by an horizontal force P, then there will be a force of friction F in the opposite direction which prevents the motion of the body. Let the resultant of R and F is S which makes an angle θ with R. Resolving S along R and F, we have

$$S \cos \theta = R \text{ and } S \sin \theta = F$$

$\therefore$
$$\tan \theta = \frac{F}{R}$$

For the sake of equilibrium

$$R = W \text{ and } F = P.$$

If we go on increasing the pull, the force of friction goes on increasing till we arrive at a stage when the body is just on the point of moving. This stage is called as limiting equilibrium. The force of friction in this case is called as *limiting*

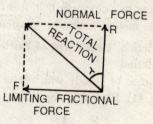

Fig. 2

friction and is maximum. The angle, which the resultant of this maximum force and normal reaction makes with the normal is called angle of friction. This is generally denoted by λ.

The *coefficient of friction* μ is the ratio of limiting friction F to the normal reaction R between two surfaces, i.e.,

$$\mu = \frac{F}{R} \qquad ...(1)$$

The angle which the resultant of limiting friction F and the normal reaction R makes with the normal is known as *angle of friction* and is denoted by λ (Fig. 2).

From Fig. 2, $\qquad \tan \lambda = \dfrac{F}{R} \qquad ...(2)$

From eq. (10 and eq. (2), we have

$$\mu = \tan \lambda \qquad ...(3)$$

4. Angle of Repose (α)

This angle is relevant to an inclined plane. If a body is placed on an inclined plane and it is just on the point of sliding down, then the angle of inclination of the plane with the horizontal (Fig. 3) is called the angle of repose (α) for the two surfaces in contact.

From Fig. 3 $\qquad F = mg \sin \alpha$

and $\qquad R = mg \cos \alpha$

$\therefore \qquad \dfrac{F}{R} = \tan \alpha = \mu \qquad ...(1)$

Again $\qquad \mu = \tan \lambda$

So $\qquad \alpha = \lambda$

Angle of repose = angle of limiting friction.

5. Laws of Friction

Various laws of friction are as under

(1) The force of friction at the point of contact of two bodies is in the direction opposite to that in which the point of contact starts moving.

(2) When the body is just on the point of moving, the force of friction is limiting. The ratio of limiting friction to normal reaction bears a constant ratio and is denoted by μ. The limiting friction is μR.

(3) The limiting friction is independent of areas in contact provided the normal reaction is unaltered.

(4) When the body starts moving, the above law of limiting friction still holds good and is independent of velocity.

6

Gravitation, Acceleration Due to Gravity and Satellites

1. Newton's Law of Gravitation

The force between any two particles having masses m_1 and m_2 separated by a distance r is an attraction acting along the line joining the particles and has the magnitude

$$F = G \frac{m_1 m_2}{r^2}$$

where G is a universal constant having the same value for all pairs of particles.

The constant G has the dimension $L^3 M^{-1} T^{-2}$ and is a scalar quantity

$$= 6.673 \times 10^{-11} \text{ N m}^2/\text{kg}^2.$$

The acceleration due to gravity g_0 at the surface of the earth $= \dfrac{GM_E}{R_E^2}$, where M_E is the mass of the earth and R_E is the radius of the earth.

The acceleration due to gravity decreases as distance 'r' from the centre of the earth increases.

The variation in g, for a small change dr in r is given by the formula $\dfrac{dg}{g} = -2\dfrac{dr}{r}$.

Value of g decreases when we go deep inside the earth also. The rate of decrease now is half of the case of going above the surface of the earth. Thus, value of 'g' is maximum, on the surface of the earth.

There is a variation in 'g' on the surface of the earth also. This is on account of (a) oval shape of the earth and (b) rotation of the earth. Value of 'g' is minimum at the equator and maximum at the poles.

Gravity. If one body is earth of mass M and radius R, then the force of attraction between the earth and body of mass m placed on the surface of the earth is called gravity and the force of gravity F_1 is given by

$$F_1 = G\frac{Mm}{R^2}$$

2. Acceleration due to Gravity

When a body falls freely, it is attracted by the earth with a force by Newton's law of gravitation. This force is called gravity. The effect of a force on the body is to produce an acceleration in it (Newton's law of motion). So the force of gravity produces an acceleration due to gravity. This is denoted by g.

According to Newton's second law of motion.

$$F_1 = mg.$$

Definitions and Formulae in Physics 43

where F_1 is the force of gravity.

$$\therefore \quad G\frac{M m}{R^2} = mg.$$

or
$$g = \frac{GM}{R^2}$$

This is relation between g and G.

3. Variation of Acceleration due to Gravity

(i) **Due to altitude.** For a body at a height h above the surface of earth, the acceleration due to gravity g_h is give by

$$g_h = g\left(1 - \frac{2h}{R}\right) \qquad ...(1)$$

where g is acceleration due to gravity on the earth and R is the radius of the earth.

Thus the value of g decreases as we go away from the surface of the earth.

(ii) **Due to depth (below earth's surface).** For a body at a depth h below the earth's surface, the acceleration due to gravity g_h'

$$g_h' \approx g\left(1 - \frac{h}{R}\right) \qquad ...(2)$$

where g is acceleration due to gravity on earth's surface and R is the radius of the earth.

Thus the value of g decreases as we go towards the centre inside earth.

(iii) **Effect of earth's rotation.** Let us assume the earth to be a sphere of radius R and mass M. Let the axis of rotation be ZOZ' as shown in Fig. (1) and its angular velocity be ω. Let us consider the case of a particle of mass m at P so

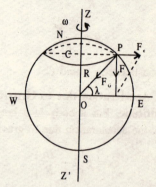

that OP makes an angle λ is the latitude of the particle. It is quite clear from the figure that the particle is moving in a circle of radius $CP = R \cos \lambda$ with angular velocity ω.

Let g' be the effective value of g due to earth's rotation, then

$$g' = g - R \omega^2 \cos^2 \lambda \qquad ...(3)$$

Thus the value of g is decreased due to earth's rotation. At equator $\lambda = 0°$ and at poles $\lambda = 90°$; therefore the decrease in g is maximum at equator and zero at poles.

(iv) **Due to shape of earth.** The earth is not perfectly spherical but it is slightly ellipsoidal. It is bulging at the equator and flattened at the poles. So the equatorial radius is more than the polar radius. As $g \propto (1/R)^2$, the value of g increases from equator to poles. The due to the shape of the earth, g is maximum at poles and minimum at equators.

4. Gravitational Field Strength (Gravitational Intensity), Potential and Potential Energy

(a) **Gravitational field strength.** The space surrounding a particle where in gravitational force can be experienced is known as the gravitational field. The gravitational intensity I is defined as the force experienced by a unit mass placed in the gravitational field of mass M, i.e.,

$$I = GM/r^2$$

where r is the distance of unit mass from M.

(b) **Gravitational potential.** The gravitational potential V at a point in the gravitational field is defined as the workdone in taking a unit mass from infinity to that point. This is given by

$$V = -\frac{GM}{r}$$

It should be remembered that $I = -(dV/dr)$

(c) **Gravitational potential energy.** The work obtained in bringing a body from infinity to a point in the gravitational field is known as the gravitational potential energy of the body at that point. The gravitational potential energy at infinity is assumed to be zero. Since work is obtained in bringing the body from infinity into the gravitational field, so the gravitational potential energy is always negative. The gravitational potential energy of a particle of mass m is given by

$$U = -\frac{GMm}{r}$$

5. Escape Velocity

It is the least velocity needed to throw a body away from the surface of the earth so that it may not return.

It can be calculated that the work necessary to take a body of mass m to a distance where earth's attraction is negligible.

$$W = \frac{GMm}{R}$$

The work may be done by a body of equal kinetic energy. If v_e be the velocity of body corresponding to this work done, then kinetic energy

$$E = \frac{1}{2} m v_e^2$$

$\therefore \qquad \frac{1}{2} m v_e^2 = \frac{GMm}{R}$

or $\qquad v_e^2 = \frac{2GM}{R} = \frac{2gR^2}{R} = 2gR$

$$v_e = \sqrt{(2gR)}$$

where v_e is known as escape velocity

$$v_e = \sqrt{(2 \times 9.8 \times 6.4 \times 10^6)}$$

$$= 11.2 \times 10^3 \text{ m/sec.}$$

6. Planets and Satellites

(a) **Planets.** Each one of the nine members of our solar system, is called a planet. The nine planets are Mercury, Venus, Earth, Mars, Jupiter, Saturn, Uranus, Neptune and Pluto. They revolve round the sun in their own orbits.

(b) **Satellites.** A satellite is any body which is revolving around a large body under the influence of the latter. For example, moon is a satellite of earth. This is natural satellite. On the other hand, there may be artificial satellites. For example, Aryabhatta, Rohini, Inset B are artificial satellites. The

Definitions and Formulae in Physics

artificial satellites are put in orbits round the earth with the help of multistage rocket. The multistage rocket carries the satellite to the required vertical height and then gives it appropriate horizontal velocity required for stable orbit around the earth.

(c) **Communication satellite.** This is an artificial satellite of earth which appears stationary to an observer on earth's surface. Such a satellite is called geo-static satellite or geo-synchronous satellite. For such a satellite, the following conditions are essential.

(i) The orbit of the satellite must be circular and in the equatorial plane of the earth.

(ii) The period of rotation of the satellite must be equal to the period of rotation of the earth about its axis.

(iii) The angular velocity of satellite must be in the same direction as angular velocity of rotation of earth.

Inset B is such a satellite in India.

7. Motion of Planets and Satellites

All planets move in elliptical orbits, the sun being at one focus. We may for simplicity consider the orbits to be very nearly circular, with the sun at the centre. The centripetal force required for the circular motions is provided by the gravitational attractions. Even though both the bodies revolve around their common centre of mass, if one is of much greater mass than the other, the heavier body may be considered to be at rest. This may be applied also to the case of the earth and the artificial satellites.

If M is the mass of the heavy body at the centre of the circle of radius r, and m the mass of the lighter body,

$$\frac{GM}{r^2} = m\omega^2 r \text{ or } \frac{GM}{r^2} = \omega^2 r$$

$$\frac{GM}{r^3} = \omega^2 = \frac{4\pi^2}{T^2}$$

From this, we may deduce the period of the planet or the satellite.

$$T^2 = \frac{4\pi^2}{GM} r^3$$

8. Kepler's Laws

I. The path of a planet is an elliptical orbit, with the sun at one of its foci.

II. The radius vector drawn from the sun to a planet sweeps out equal areas in equal intervals of time.

III. The square of the planet's period is proportional to the cube of the semi major axis of its orbit.

9. Orbital Velocity

If a satellite is to keep moving in a circular orbit round the earth, at a distance h from the surface of the earth,

$$\frac{GM}{(R_E + h)^2} = \frac{mv_0^2}{(R_E + h)}$$

where R_E is the radius of the earth

$$v_0 = \sqrt{\frac{GM}{(R_E + h)}}$$

if $\qquad h < R_E, v_0 = \sqrt{gR_E}$

Definitions and Formulae in Physics

10. Escape Velocity

The gravitational potential energy of a particle of mass m at the surface of the earth

$$= -G\frac{M_E m}{R_E}$$

The amount of work required to move a body from the earth to infinity is given by the formula

$$\frac{G_E M m}{R_E}$$

This is about 0.6×10^7 joules/kg. If we could provide a projectile more than this energy at the surface of the earth, it would escape from the earth. The critical initial speed v_0 called escape velocity is given by the formula

$$\frac{1}{2}mv_0^2 = \frac{GM_E m}{R_E}$$

$$v_0 = \sqrt{\frac{2GM_E}{R_E}} = 11.2 \text{ km/second}$$

11. Synchronous Satellite

This is an artificial satellite whose period is 24 hours, so that it is always vertically over the same place on the earth. For this to be possible.

$$T = 2\pi\sqrt{\frac{(R_E + h)^3}{GM_E}} = 86400 \text{ seconds.}$$

'h' may be calculated. It is about 22,700 miles above the earth's surface.

12. Jet

A jet is a stream of liquid or gas coming out through a narrow hole at high pressure.

Jet Propulsion is the motion of a body due to reaction of jet. It depends on Newtons third law and law of conservation of momentum.

7

Uniform Circular Motion

1. Circular Motion

When a particle tied with a string is whirled in a horizontal plane and the other end of the string is kept fixed at a point, the particle executes circular motion with centre at the fixed point.

2. Centripetal Force

Centripetal force is the force needed to give circular motion to the particle or to a body. When a particle moves in a circle of radius r, its speed is constant but the direction of velocity changes continuously. Thus the particle experiences an acceleration. The magnitude of the acceleration is v^2/r is always directed towards the centre of the circle. According the Newton's law, the particle must be under the action of an external force so as to have this acceleration. The external force that must be imposed on the body, in order to make it move with uniform speed around the circle is known as centripetal force.

As $$F = ma = m\frac{v^2}{r} \qquad \ldots(1)$$

The direction of F at any instant must be radially inward.

In terms of the angular velocity of the body, ω

$$F = m\omega^2 r. \qquad ...(2)$$

3. Conical Pendulum

A conical pendulum consists of a string AB (Fig. 1) whose upper end is fixed at A and other end B is tied with a bob.

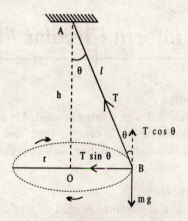

Fig. 1.

When the bob is drawn aside and is given a horizontal push, let it describe a horizontal circle with constant angular speed ω in such a way that AB makes a constant angle θ with the vertical. As the string traces the surface of a cone, it is known as conical pendulum.

The time period is given by

$$\therefore \qquad T = 2\pi \sqrt{\left(\frac{h}{g}\right)}$$

4. Banking of Tracks

When a vehicle moves round a curve on the road with excessive speed, then there is a tendency for the vehicle to overturn outwards. To avoid this, the road is given a slope rising outwards. The outer wheel of the vehicle is now raised. This is called as banking.

$$\tan \theta = \frac{v^2}{r \cdot g}$$

where θ is the angle of banking and r is the radius of curved path.

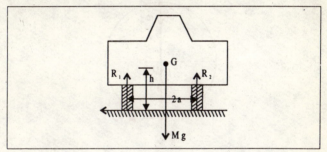

Fig. 2.

(i) the weight mg of the car acts downwards through centre of gravity G.

(ii) the normal reactions of the ground R_1 and R_2 on the inner and outer wheels respectively. These act vertically upward.

(iii) the force of friction F between wheels and ground towards the centre of the turn.

Let the radius of circular path be r and the speed of the car be v.

The maximum speed without overturning is give by

$$v = \sqrt{\left(\frac{g\,r\,a}{h}\right)}$$

5. Motion in Vertical Circle

Taking up the case of a body A of mass m tied to a string of length l whose other end O is fixed. If the body is projected with a velocity u at right angle to OA. And if the velocity is small, the body and string execute oscillations.

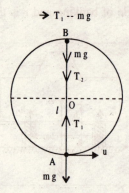

Fig. 3.

When the velocity is large, the body describes a circle. The velocity of the body and tension in the string so that the body completes a circle in vertical plane can be found as under.

First of all calculate the tensions in the string at positions A and B (Fig. 3). At point A, the resultant force acting on the body is given by

$$T_1 = mg$$

This is equal to the centripetal force

$$T_1 = m\left(g + \frac{u^2}{l}\right)$$

This tension is always positive i.e., greater than zero. Thus string will be tight in this position.

The resultant force acting on the body at B is given by

$$T_2 + mg = \frac{mu_1^2}{l} \quad \text{(where } u_1 \text{ is the velocity at } B\text{)}$$

$\therefore$
$$T_2 = m\left(\frac{u_1^2}{l} - g\right)$$

If, T_2 is greater than zero, the string will be tight and if T_2 is negative the string becomes loose and does not perform circular motion. Thus, the condition for the body to complete a circle is that the tension in the string should be greater than zero.

Let us consider a more general case in which the position of the body is at P at any instant as shown in Fig. (4). The different force are also shown in the same figure. Let $AD = h$ and $\angle DOP = \theta$.

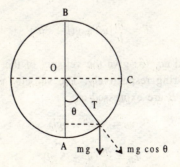

Fig. 4.

The velocity v of the particle at P is give by

$$v^2 = u^2 - 2gh \qquad \ldots(1)$$

The centripetal force mv^2/l is equal to $(T - mg \cos \theta)$. Hence

$$T - mg \cos \theta = \frac{mv^2}{l} \qquad \ldots(2)$$

From Fig., $\quad \cos \theta = \dfrac{OD}{OP} = \dfrac{l-h}{l} \qquad \ldots(3)$

Substituting the value of $\cos \theta$ from eq. (3) in eq. (2), we get

$$T - mg\left(\frac{l-h}{l}\right) = \frac{mv^2}{l}$$

or $\quad T = \dfrac{mv^2}{l} + \dfrac{mg}{l}(l-h) \qquad \ldots(4)$

Substituting the value of v in eq. (4) from eq. (1), we get

$$T = \frac{m(u^2 - 2gh)}{l} + \frac{mg}{l}(l-h)$$

or $\quad T = \dfrac{m}{l}[u^2 + g(l - 3h)] \qquad \ldots(5)$

Eq. (1) and eq. (5) give the velocity of the body and tension in the string respectively. The velocity v_B and tension T_B at a point B are expressed as

$$v_B = \sqrt{(u^2 - 4gl)} \qquad \ldots(6)$$

and $\quad T_B = \dfrac{m}{l}(u^2 - 5gl) \qquad \ldots(7)$

Definitions and Formulae in Physics

Here we consider the following cases :

Case 1. Condition to perform complete revolution

The condition for the body to complete revolution is that the tension in the string at B should be greater than zero i.e.,

$$u^2 - 5\,gl > 0 \text{ or } u^2 > 5\,gl \text{ or } u > \sqrt{(5\,gl)}.$$

Case 2. Condition of oscillation

If $u < \sqrt{(5gl)}$, the particle will either oscillate about the lowest point A or will leave the circular path.

Let the velocity vanish at some height h_1 then from eq. (1).

$$0 = u^2 - 2\,hg_1 \therefore h_1 = u^2/2g.$$

Again let the tension of the string vanish at some height h_2 then from eq. (5)

$$u^2 + g(l - 3\,h_2 = 0 \therefore h_2 = \frac{u^2 + gl}{3\,g}$$

The condition of oscillation of the body is that the velocity should vanish earlier than tension i.e., $h_1 > h_2$

$$\therefore \qquad \frac{u^2}{2\,g} < \frac{u^2 + gl}{3\,g}$$

$$3u^2 < 2u^2 + 2gl,\ u^2 < 2gl,\ u < \sqrt{(2gl)}$$

so, when $u < \sqrt{(2gl)}$, the body will oscillate about the lowest point. if $u = \sqrt{(2gl)}$, the arc of oscillation is semi-circle.

Case 3. Condition for the body to leave the circular path

This occurs when the string becomes slack i.e., tension vanishes earlier than velocity, i.e.,

$$\frac{u^2 + gl}{3g} < \frac{u^2}{2g} \text{ or } u^2 > 2gl \therefore u > \sqrt{(2gl)}$$

so if $\sqrt{(2gl)} < u < \sqrt{(5gl)}$, the particle will leave circular path somewhere between C and B.

8

Rotational Motion

1. Rotational Motion

In rotational motion of a rigid body about a fixed axis, the *angular displacement* is expressed in degree or radians.

The angular displacement

$$= \frac{\text{Linear displacement along a circle}}{\text{Radius of the circle}}$$

$\theta = \dfrac{s}{r}$ radians θ = angular displacement

s = linear displacement
r = radius of the circle

2. Angular Velocity

It is defined as the rate of change of angular displacement with time.

So $$\omega = \frac{\theta}{t} = \frac{s}{tr} = \frac{v}{t}$$

where ω = angular velocity

v = linear velocity

∴ Angular velocity

$$= \frac{\text{Linear velocity}}{\text{Radius of the circle}}$$

The unit of angular velocity is radian/sec.

3. Angular Acceleration

α is the time rate of change of its angular velocity. Its unit is radian/sec^2.

We get the following relations :

linear displacement $s = r\theta$
linear velocity $v = rw$
linear acceleration $f = ra$

4. Moment of Inertia

Moment of inertia of a body is a measure of the resistance a body offers to any change in its angular velocity. If a body is made of masses, m_1, m_2, m_3... at respective distances r_1, r_2, r_3... etc., from an axis, then its moment of inertia about the axis is

$$I = m_1 r_1^2 + m_2 r_2^2 + m_3 r_3^2 + ... = \Sigma m r^2$$

Moments of Inertia of some symmetrical bodies :

For a small mass m at distance r from the axis of rotation

$$I = mr^2$$

For a uniform solid disc or cylinder of total mass m and radius r about an axis passing through its centre and perpendicular to its face

$$I = \tfrac{1}{2}mr^2$$

For a thin uniform rod of mass m and length l about an axis through its middle point and at right angles to its length

$$I = \tfrac{1}{2} m l^2$$

For a uniform solid sphere of mass m and radius r about any diameter.

$$I = \tfrac{2}{5} m r^2$$

5. Kinetic Energy of a Rotating Body
Rotational KE

= ½ Moment of inertia × square of the angular velocity

= $\tfrac{1}{2} \omega^2$.

6. Angular Momentum
= Moment of inertia × angular velocity

= $I \times \omega$.

Angular momentum about an axis is the product of moment of Inertia (I) and the angular velocity i.e., $J = I \times \omega$.

The units of angular momentum are kg-metre^2/sec or joule-sec.

7. Law of Conservation of Angular Momentum
It states that total angular momentum of a system remains constant provided no external force acts on it.

$$I \times \omega = \text{constant}.$$

8. The Moment of Force
About an axis is the effectiveness of the force in producing rotation about that axis. It is measured by the product of the force and the perpendicular distance from the axis of rotation to the line of action of the force. It is a vector quantity.

∴ *Moment* = force × perpendicular distance from axis to the line of action of the force, the unit of moment in MKS system in $nt \times m$.

9. A Couple Constitutes

Two equal and oppositely directed parallel forces not in the same straight line.

Moment of a couple is measured by the product of one of the forces and the perpendicular distance between them.

10. Relativity

(i) *Variation of mass* is predicted by the special theory of relativity. The mass of a moving body is greater than its mass at rest (called the rest-mass).

$$m = \frac{m_0}{\sqrt{1 - v^2/c^2}}$$

where m is the mass when moving with a velocity v. m_0 is the rest mass and c is the velocity of light in vacuum.

(ii) *Equivalence of mass and energy.*

When mass m is converted completely into energy W, we have the relation

$= mc^2$ where c is the velocity of light in vacuum.

or W (joule) $= m$ (kg) $\times c$ (metre/sec) $\times c$ (metre/sec).

(iii) *Relativistic contraction of length*

The length of an object in motion with respect to an observer is measured by the observer to be shorter than when it is at rest with respect to him. Thus

$$L = L_0\sqrt{1 - (v^2/c^2)}$$

Definitions and Formulae in Physics

where L_0 is the length when the object is at rest L is its length when it is moving with velocity v and c is the velocity of light in vacuum.

11. Matter Waves

The equivalence of mass and energy given by $W = mc^2$ indicates that a photon of energy hf may be considered as a particle of mass m given by $hf = mc^2$.

$\therefore$
$$m = \frac{hf}{c^2} = h \times \frac{c}{\lambda} \times \frac{1}{c^2} = \frac{h}{\lambda}$$

or
$$\lambda = \frac{h}{mc} = \frac{h}{\text{momentum of the photon}}$$

Similarly a moving particle exhibits wave properties such that the wavelength λ associated with its mass m and velocity v is given by

$$\lambda = \frac{h}{\text{momentum of the particle}}$$

$$= \frac{h}{mv}$$

These waves are called **matter waves**.

9

Elasticity

1. Basic Terms

Elasticity. It is a property by virtue of which a body regains its original size and shape after a deformation when the deforming forces have removed.

Stress It is the force applied per unit area which produces deformation. It is expressed as nt/m^2.

Strain. It is the ratio of the change in some dimension of the body to the total value of the dimension in which change has taken place.

Elastic limit. It is the minimum stress that will produce a permanent set.

2. Hooke's Law

It states that within the elastic limit, the ratio of the stress to the strain produced is a constant. This constant is called *modulus of elasticity*. It is expressed in nt/m^2

I. Young's modulus

Definitions and Formulae in Physics

$$Y = \frac{\text{longitudinal stress}}{\text{longitudinal strain}}$$

II. *Bulk modulus*

$$B = \frac{\text{volume stress}}{\text{volume strain}}$$

III. *Rigidity (shape) modulus*

$$n = \frac{\text{shearing stress}}{\text{shearing strain}}$$

Young's modulus for a wire of radius r, length L is given by

$$Y = \frac{Mg}{\pi r^2} \div \frac{l}{L} = \frac{MgL}{\pi r^2 l}$$

where
- M is mass suspended
- l is increase in length of the wire
- g is acceleration due to gravity

3. Poison's Ratio

Poisson's ratio σ is defined as the ratio of fractional change in diameter to the fractional change in length. Thus

$$\sigma = (\Delta r/r)/(\Delta l/l) = \Delta r \cdot l/\Delta l \cdot r$$

Here r = radius of the wire.

4. Work done (Energy Stored) in Stretching a Wire

Suppose a wire of length L and area of cross section A is stretched through a distance l by applying a force F along its length. Then

$$\text{Stress} = F/A \text{ and strain} = l/L$$

Work done per unit volume = (1/2) stress × strain.

This workdone is stored as the strain energy is the wire.

5. Thermal Stresses

If the ends of road are rigidly fixed and its temperature is changed, then compressive stresses (thermal stresses) are set up in the rod. The thermal stresses may be beyond the breaking strength of the rod. Here in the design of any structure some provision is made for such expansion. The thermal stress set up in the rod which is not free to expand or contract is given by

$$\therefore \quad \text{Stress in the rod} = \frac{F}{A} = Y \alpha \Delta T,$$

Where Y is young's modulus, α, is the coefficient of linear expansion and ΔT is fall in temperature.

10

Fluid Motion and Pressure in a Fluid

1. Fluid Motion

When the fluid velocity $\vec{V}$ at any given point is constant in time, the fluid motion is said to be *steady*. That is, at any given point in a steady flow, the velocity of each passing fluid particle is always the same.

In *non-steady* flows as in a tidal bore, then velocities V are a function of time.

In case of *turbulent* flow, the velocities vary erratically from point to point and from time to time.

Fluid flow is irrotational, if the element of fluid at each point has no net angular velocity about that point.

Fluid flow can be compressible or incompressible. Liquids can be considered incompressible.

Fluid flow can be *viscous* or *nonviscous*. Viscosity in fluid motion is the analogue of friction in the motion of solids.

2. Dynamics of Steady, Irrotational, Incompressible and Nonviscous Flow

Steam Line. In steady flow the velocity $\vec{V}$ at a given point is constant in time. A stream line is a curve which is parallel to the velocity of fluid particle at every point. No two stream lines an cross one another. The pattern of stream lines is stationary with time.

Tube of flow. A bundle of a finite number of stream lines forms a tube of flow. No fluid can cross the boundaries of a tube of flow. The fluid that enters at one end leaves at the other end.

3. Equation of Continuity

If A is the area of cross-section of a tube at right angles to the stream lines at a point, ρ the density of the fluid at that point, and v the velocity of flow.

$$\rho A v = \text{constant}$$

If the fluid is incompressible.

$$Av = \text{constant or } A_1 v_1 = A_2 v_2$$

The product Av is called the volume flux or flow rate.

Density. *The density ρ of a substance is the ratio of its mass to its volume, i.e.,*

$$\rho = (M/V)$$

where M is the mass of the body and V is volume.

Relative density. The relative density is the ratio of the density of the substance to the density of water at 4°C.

$$\text{relative density} = \frac{\text{density of the substance}}{\text{density of water at 4°C}}$$

$$= \frac{\text{Mass of any volume of substance}}{\text{Mass of the equal volume of water at 4°C}}$$

The relative density has no unit because it is pure ratio.

3. Archimede's Principle

According to Archimede's principle, when any body (totally or partially) is immersed in a liquid it appears to lose part of its weight and the apparent loss of weight is equal to the weight of liquid displaced.

Let a body of mass M and volume V is suspended in a liquid of density ρ by a string, then

Apparent weight of the body in liquid $= (M - V\rho)g$

Loss of weight in liquid $= (Mg - (M - v\rho)g)$

$\qquad\qquad\qquad\qquad = V\rho g$

$\qquad\qquad\qquad\qquad$ = wt of equal volume of displaced liquid

According to Archimede's principle relative density

$$= \frac{\text{weight of the substance in air}}{\text{weight of equal volume in water}}$$

$$= \frac{\text{weight of the substance in air}}{\text{loss of weight in water}}$$

4. Laws of Floatation

When a body floats in a liquid, then the equilibrium is governed by the two forces

(i) The weight of the body acts vertically downwards through centre of gravity of the body.

(ii) The resultant upwards thrust exerted on the body by the liquid i.e., *force of buoyancy* acts vertically upwards through the centre of gravity of displaced liquid, i.e., centre of buoyancy.

When the body floats freely, we have

(1) The weight of the body is equal to the weight of the displaced liquid.

(2) The centre of gravity of the body and centre of gravity of the displaced liquid are in the same vertical line.

These are known as laws of floatation.

Thus the essential condition for a body to float in a liquid is that the weight of the liquid displaced by the body should be equal to the weight of the body itself.

5. Liquid Pressure

A liquid contained in a vessel exerts a force on the bottom of the vessel and on the sides of the vessel. This force is normal to the surface. The pressure is defined as force per unit area. In C.G.S. system the unit of pressure is dyne/cm^2 while in S.I. unit it is newton/m^2.

The pressure due to a liquid column of height h is given by

$$P = h\rho g$$

where ρ is the density of the liquid and g is acceleration due to gravity.

The thrust is defined as

Thrust = Pressure × Area = $h\rho g \times A$.

6. Atmospheric Pressure

Like liquid, air also exerts pressure which is known as atmospheric pressure. If h be the height of the atmospheric air, the atmospheric pressure P is

Definitions and Formulae in Physics

$$P = hdg$$

where d is the density of air.

Characteristics of Liquid Pressure

Important characteristics of a liquid pressure are as under:

(1) The pressure at a point is same in all directions.

(2) The force exerted by a liquid is perpendicular to any surface with which the liquid is in contact.

(3) An increase in pressure at any point in a confined liquid produces an equal increase in pressure at every other point in the liquid. This is called Pascal's law.

7. Boyle's Law

Boyle's law states that the volume of a given mass of a gas varies inversely as the pressure of the gas if the temperature is kept constant. Thus

$$P \propto (1/V) \text{ or } PV = \text{constant.}$$

If P_1 and V_1 be the pressure and volume of a gas under one condition and P_2, V_2 be the similar quantities under other condition, then

$$P_1V_1 = P_1V_2$$

8. Pascal's Law

Pascal's law states that the pressure exerted any where upon a mass of a confined liquid is transmitted by the liquid equally in all directions so as to act with undiminished force per unit area and at right angles to the surface exposed to the liquid.

9. Bernoulli's Equation

Consider a tube of flow of a fluid, the flow being steady, incompressible and non-viscous. At any given point of the tube, let the mean height of the tube above a reference level be y. Let p be the pressure, ρ the density, and v the velocity of flow at the point. Then

$$p + \frac{1}{2}\rho v^2 + \rho g y = \text{constant for the tube of flow.}$$

This is Bernoulli's equation. This principle and the equation of continuity are used in the Venturi metre and the point tube for measurement of fluid flow.

10. Application of Bernoulli's Theorem

Calculation of velocity of efflux, volume coming out per unit time through a hole in a tank filled with liquid, volume flowing per second in a venturimeter, action of Pilot's tube swinging of a cricket ball, working of an atomiser, blowing off the roof of building in a tornado, etc., are typical examples where Bernoulli's theorem is used.

11. Velocity of Efflux

Velocity of efflux for flow of a liquid through a hole in a tank filled to a height 'h' above the hole, comes out to be $v\sqrt{2gh}$. If 'a' be the area of cross section of the hole, theoretically, the volume of the liquid coming out per unit time should be $V = a \times \sqrt{2gh}$. However, practically it comes out to be $V = 0.62\, a\sqrt{2gh}$. The phenomenon is known as venacontracts. The reason for this is that whole liquid does not fall incident normal to the cross section of the hole.

12. Venturimeter

Basically it is a horizontal pipe with different cross sections a_1 and a_2, in which a liquid is allowed to flow. If the pressures at these cross sections are p_1 and p_2, using equation of continuity of liquid flow and Bernoulli's theorem, one can calculate, the volume flowing per second as,

$$V = a_1 a_2 \sqrt{\frac{2(p_1 - p_2)}{\rho(a_1^2 - a_2^2)}}$$

Thus, it can be used for calculating the rate of flow of a liquid.

13. Dynamic Lift

Dynamic lift is the force that acts on a body, such as an airplane wing, a hqdrofoil, or a spinning ball, by virtue of its motion through a fluid. This is different from *static lift* which is the buoyant force that acts on an object.

14. Torricell's Theorem

It states that the velocity of efflux of a liquid through an orifice is equal to that which a body would acquire in falling freely from the free surface of the liquid to the orifice. The velocity is given by

$$v = \sqrt{(2gh)},$$

where h is the height of the orifice below the free surface of the liquid.

11

Surface Tension and Viscosity

1. Surface Tension

When a molecule is brought from the interior of the liquid to the surface, work is done against the downward cohesive force acting on it. Thus the potential energy of the molecules in the surface film is greater than those inside the liquid. Since potential energy tends to be a minimum, the surface tends to have the least area. The potential energy per unit area of the surface film is called *surface energy*.

The surface tension T in a liquid may be defined as the amount of work done in increasing the surface area of the liquid film by unity.

2. Excess of Pressure

Excess of pressure inside a liquid drop which is spherical and of radius r is given by the formula

$$p = \frac{2T}{r}$$

Excess of pressure inside a soap bubble

$$= \frac{4T}{r}$$

Work done in blowing a bubble of radius $r = 8\pi r^2 T$.

3. Capillary Rise

When one end of a capillary tube is dipped in a liquid of surface tension T, the liquid rises to a height h in the capillary.

$T = \dfrac{rh\rho g}{2}$, where r is the radius of the capillary and ρ is the density of liquid.

If the tube is of insufficient length, that is its height is less than h, then the meniscus assumes a radius of curvature R given by the formula

$Rl = \dfrac{2T}{\rho g}$, where l is the length of the tube (less than h).

4. Factors Affecting Surface Tension

(i) *Contamination*

(ii) *Dissolved substances* : (a) When the substance is highly soluble, surface tension increases (b) when the substance is only feebly soluble (c) When the substance is only feebly soluble, surface tension decreases.

(iii) *Change of temperature* : Surface tension of unassociated liquids decreases with the rise of temperature.

5. Viscosity

The coefficient of viscosity of a liquid may be defined as the tangential force required per unit area to maintain a unit velocity gradient.

If this tangential force be unity the coefficient of viscosity is unity and is called Poise.

6. Critical Velocity

The critical velocity V_c is given by the relation

$$V_c = \frac{k\eta}{\rho r},$$

where η is the coefficient of viscosity, ρ the density and r the radius of the tube. The constant k is called *Reynold's number*. Its velocity gives the upper limit of steady flow.

7. Poiseuille's Equation

If P is the difference of pressure between the two ends of a capillary tube of length L, and radius of cross-section r, the volume of steady flow of liquid through the tube per second is given by the formula

$$V = \frac{\pi P r^4}{8L\eta},$$ where η is the coefficient of viscosity.

8. Stoke's Formula

Stoke showed that the retardation F due to the viscous drag for a spherical body of radius r, moving with velocity v in a medium whose coefficient of viscosity is η is given by

$$F = 6\pi v r \eta$$

9. Surface Energy

Surface energy is the potential energy of the surface of a liquid. If the surface is extended by a certain amount of area, the surface energy possessed by a body is this increase in area times the surface tension. Thus, surface tension may also

Definitions and Formulae in Physics

defined as the surface energy per unit area. By this definition, the units of surface tension will be Joules/metre² or ergs/cm².

10. Excess Pressure Inside Surface.

(i) Excess pressure inside a liquid surface is given by

$$p = 2\sigma/r$$

where σ is the surface tension and r is the radius of the bubble.

(ii) Excess pressure inside a soap bubble is given by

$$p = (4\sigma/r)$$

as a soap bubble has two surfaces internal and external.

For hemispherical shapes in either case, it would be half of what has been given above.

11. Angle of Contact, Shape of the Meniscus, etc.

Some liquids have concave and some others convex surfaces. For concave surfaces, adhesive forces between the liquid molecules and those of the container are stronger than the cohesive forces amongst molecules of the liquid. Similarly, for convex surfaces, cohesive forces are stronger than the adhesive forces.

12. Capillarity Action

As a capillary is dipped in a liquid, if the liquid exhibits concave surface, liquid rises in it upto a certain height and if the liquid exhibits convex surface, liquid depresses into it upto a certain depth give by

$$h = \frac{2\sigma \cos\theta}{r\rho g}.$$

Here σ is the surface tension, θ is the angle of contact, r is the radius of the capillary and ρ is the density of the liquid.

13. Angle of Contact

It is the angle between the tangent to the liquid surface at the interface and the solid surface inside the liquid. For concave surfaces, it is acute and for convex surfaces, it is obtuse.

14. Applications of Surface Tension

The following are some examples of applications of surface tension :

(i) Working of fountain pen

(ii) Seepage of water through brick walls

(iii) Spherical shape of rain drops

(iv) Flat convex shape of mercury drops

15. Variation of Surface Tension with Temperature

The cause of surface tension is intermolecular forces. On rise in temperature, on account of expansion, intermolecular distances increase, resulting in decrease in these forces and hence decrease in surface tension with rise in temperature.

16. Streamline and Turbulent Flow

If the rate of flow of a liquid is not very large, it takes place in the form of layers with regular gradation in their velocities. Such a flow is called Streamline Flow. On the other hand, if the rate of flow is quite much, irregularities start occurring in the flow and it no longer takes place in the form of layers with regular gradation in velocities. This type of flow is called Turbulent Flow.

17. Reynold's Number and Critical Velocity

Critical velocity is that velocity beyond which turbulence starts. It is given by

$$V_c = \frac{K\eta}{\rho D}$$

where V_c is the critical velocity, η is the coefficient of viscosity, ρ is the density of the liquid and D is the diameter of the tube. K is the dimensionless constant called Reynold's number and for most tubes it lies between 500 to 1000.

12

Thermometry and Thermal Expansion

1. Heat
Heat is the agent which gives the sensation of hotness or coldness of the body.

2. Temperature
The temperature of a body is that physical quantity which indicates how much hot or cold the body is. So the *temperature of a body is measure of its hotness or coldness.*

3. Zeroth Law of Thermodynamics
If A and B are in thermal equilibrium with a third body C, then A and B are in thermal equilibrium with each other also.

Or

There exists a scalar quantity called temperature, which is a property of all thermodynamic systems (in equilibrium states) such that temperature equality is a necessary and sufficient condition for thermal equilibrium.

4. Measurement of Temperature

The standard fixed-point in thermometry is the triple point of water which is arbitrarily assigned a value of 273.16 k. At this point, ice, liquid water and water vapour coexist in equilibrium. This state can be achieved only at a definite pressure, and is unique.

Common Scales of Temperature

There are three commonly used scales of temperature :

(i) **Centigrade or Celesius scale.** In this scale lower fixed point i.e., temp. of ice is taken $0°$ and upper fixed point i.e. temp. of boiling water is taken as $100°$.

(ii) **Fahrenheit scale.** In this scale lower fixed point is taken as $32°$ and upper fixed point is taken as $212°$.

(iii) **Rheumer scale.** Here lower fixed point is $0°$ while upper fixed point is $80°$.

Relation for their interconversion is

$$\frac{C}{100} = \frac{R}{80} = \frac{F - 32}{180}$$

5. Thermometre

The temperature of a body is expressed as a number, on a scale of temperature. *The instrument which deals with the measurement of temperature is known as thermometre.*

Thermometric properties of different substances are used in the construction of thermometres. The corresponding thermometers are

(i) Length of mercury column in a glass capillary (mercury thermometer)

(ii) Pressure of gas at constant volume (constant volume gas thermometre)

(iii) Electrical resistance of a metal wire (Platinum resistance thermometre)

(iv) Thermo E.M.F. (Thermo-couple thermometre)

The unknown temperature t on a scale by utilising a general property X of the substance is given by

$$t = \left(\frac{X_t - X_0}{X_{100} - X_0}\right) 100°C$$

For example, in a mercury scale

$$t_{mercury} = \left(\frac{l_t - l_0}{l_{100} - l_0}\right) 100°C$$

where l_0, and l_{100}, and l_t be the lengths of mercury column at 0°C, 100°C and at unknown temperature t°C respectively.

Constant volume gas scale

$$t_{gas} = \left(\frac{P_t - P_0}{P_{100} - P_0}\right) 100°C$$

where P_0, P_{100} and P_t be the pressures of a given mass of the gas (at constant volume) at 0°C, 100°C and an unknown temperature t°C.

$$t_{platinum} = \left(\frac{R_t - R_0}{R_{100} - R_0}\right) 100°C$$

where R_0, R_{100} and R_t be the resistances of platinum wire at 0°C, 100°C and unknown temperature t°C.

Thus these are four types of thermometres usually put to use. These are

(i) **Gas thermometre.** Here expansion of gas under constant pressure or increase in pressure for constant volume with temp, rise is used for measurement of temperature.

(ii) **Liquid thermometre.** This makes use of expansion of liquid with rise in temperature.

(iii) **Resistance thermometre.** This utilises the property of increase in resistance with rise in temperature.

(iv) **Thermoelectric thermometre.** This thermometer makes use of Seebeck Thermoelectric Effect for temperature measurement.

Besides the above, for low temperature range we make use of Vapour Pressure Thermometre and for very high temperature range we use Radiation Pyrometers.

6. Expansion of Solids

A solid when heated, may blow expansion in length, area and volume. Correspondingly, there are three coefficients of expansion viz., coefficient of linear expansion, coefficient of superficial expansion and coefficient of cubical expansion. These coefficients can be defined as under.

(i) **Coefficient of linear expansion.** The coefficient of linear expansion is the increase in length per unit length per unit degree rise in temperature. This is denoted by α. Thus

$$\alpha = \frac{L_t - L_0}{L_0 t}$$

where, L_t = length of the rod at $t°C$, L_0 = length of the rod at $0°C$ and t is rise in temperature.

If a rod of length L_0 at $0°C$ is heated to a temperature $t°C$, then its new length L_t is given by

$$L_t = L_0 (1 + \alpha t)$$

(ii) **Coefficient of superficial expansion.** The coefficient of superficial expansion is the increase in area per unit area per degree in temperature. This is denoted by β. Thus

$$\beta = \frac{A_t - A_0}{A_0 t}$$

where A_t = surface area at $t°C$, A_0 = surface area at $0°C$ and t = rise in temperature. Further $A_t = A_0 (1 + \beta t)$.

(iii) **Coefficient of cubical expansion.** The coefficient of cubical expansion or volume expansion is the increase in volume per unit volume per unit degree rise in temperature. This is denoted by γ. Thus

$$\gamma = \frac{V_t - V_0}{V_0 t}$$

where V_t = volume at $t°C$, V_0 = volume at $0°C$

and t = rise in temperature.

Further $V_t = V_0 (1 + \gamma t)$

Relation between α, β and γ

It can be shown that $\alpha : \beta : \gamma :: 1 : 2 : 3$

i.e., $\beta = 2\alpha$ and $\gamma = 3\alpha$

Variation of density with temperature

If d_0 be the density of a solid at $0°C$, d_t, the density at $t°C$, then

$$d_0 = d_t (1 + \gamma t)$$

where γ is the coefficient of volume expansion.

7. Real and Apparent Expansions of Liquids

This shows that the density of a substance, in general, decreases with the increase in temperature.

The coeffient of apparent expansion of a liquid is the ratio of the observed increase in volume per degree of temperature to the original volume of the liquid. This is denoted by γ_a

$$\therefore \quad \gamma_a = \frac{\text{observed increase in volume}}{\text{original volume} \times \text{rise in temperature}}$$

The coefficient of real expansion of a liquid is the ratio of real increase in volume per degree rise of temperature to the original volume of the liquid. This is denoted by γ_r.

$$\therefore \quad \gamma_r = \frac{\text{real increase in volume}}{\text{original volume} \times \text{rise in temperature}}$$

It may be remembered that

$$\gamma_r = \gamma_a + \gamma_g$$

where γ_g is the coefficient of cubical expansion of the material of the containing vessel.

γ_a can be measured by a weight thermometer using the formula given below

$$\gamma_a = \frac{\text{mass of the liquid expelled}}{\text{mass of the liquid left at higher temp.} \times \text{rise of temp.}}$$

γ_r can be measured by Dulong and Petits apparatus or Regnault's apparatus using the formula given below :

$$\gamma_r = \frac{h_t - h_0}{h_0 \times t}$$

where h_t is the height of the liquid in a limb surrounded at $t°C$ and h_0, the height of liquid in a limb surrounded at $0°C$.

8. Boyle's Law

Boyle's law states that the pressure of a given mass of gas varies inversely as the volume of the gas if the temperature remains constant, i.e.,

$PV = $ constant (Temperature being constant) or $P_1V_1 = P_2V_2 = P_3V_3 = $ ......

9. Charle's Law

Charle's law states that the volume of a given mass of a gas is directly proportional to its absolute temperature if the pressure remains constant

$V \propto T$ (pressure being constant)

or
$$V_1/T_1 = V_2/T_2$$

Where V_1 and V_2 be the volume of the gas at absolute temperatures T_1 and T_2 respectively.

This law can also be stated as, volumes of a given mass of a gas at constant pressure increases by 1/273 of its value at 0°C for every degree celsius rise in temperature.

10. Gas Equation and Gas Constant

Combining Boyle's law and Charle's law, we obtain an equation known as equation of state for the ideal gas

$$\frac{P_1V_1}{T_1} = \frac{P_2V_2}{T_2} = \frac{P_3V_3}{T_3} = constant.$$

where P_1, V_1 and T_1 are pressure, volume and temperature of the ideal gas respectively in the initial state P_2, V_2 and T_2 are the corresponding quantities in final state. This is called gas

Definitions and Formulae in Physics

equation. The constant is called gas constant. It's value depends upon the nature of the gas and amount (mass) of the gas. If one gm mole of a gas is taken, then the value of the gas constant is denoted by R and is same for all gases irrespective of their natures. The value of R is 8.26×10^7 ergs per gm mole per °K or 8.26 Joule per mole per °K. The gas constant for unit mass of the gas is denoted, by r where $r = R/M$ (M being molecular mass). Thus, the gas equation of unit mass is given by

$$PV = rT$$

For a mass m of the gas, we have

$$PV = mrT = m\frac{R}{M}T = nRT$$

where $m/M = n$, the number of moles.

Some Important points which should be remembered are :

(i) The volume of a 1 gm mole of a gas at N.T.P. is always equal to 22.4 litres or 22400 c.c.

(ii) N.T.P. means normal temperature and pressure i.e., temperature is 0°C or 273°K and pressure 760 mm of mercury. Sometimes this is also denoted as S.T.P.

(iii) One gm molecule of a gas in the quantity of gas whose mass is equal to its molecular weight in gms.

11. Volume Coefficient

Consider a given mass of a gas is heated through 1°C at constant pressure. Now the ratio of the increase in volume to the original volume at 0°C is defined as volume coefficient. If V_t and V_0 be the volumes of a given mass of gas at T°C and 0°C respectively, then

$$\alpha_v = \frac{V_t - V_0}{V_0 t}$$

where α_v is called volume coefficient.

12. Pressure Coefficient

Consider a given mass of gas is heated through 1°C at constant volume. Now the ratio of the increase in pressure to the original pressure at 0°C is defined as the pressure coefficient. If P_t and P_0 be the pressure of a given mass of a gas at T°C and 0°C respectively, then

$$\alpha_p = \frac{P_t - P_0}{P_0 t}$$

where α_p is called pressure coefficient.

13. Thermal Stress

Consider the case of a rod whose ends are rigidly fixed such as to avoid expansion or contraction. When the temperature is increased or decreased, tensile or compressive stresses are set up in the rod. These are known as thermal stresses

We know that $$Y = \frac{\text{stress}}{\text{strain}} = \frac{\text{stress}}{\Delta L / L_0}$$

where ΔL is the change in length due to increase or decrease in temperature. L_0 is the original length

Now $$\text{Stress} = Y(\Delta L/L_0) = Y\left(\frac{\alpha L_0 t}{L_0}\right) \quad (\because \Delta L = \alpha L_0 t)$$

or $$\text{Stress} = Y \alpha t$$

or $$F/A = Y \alpha t \quad (\because \text{Stress} = F/A)$$

$\therefore \quad F = Y \alpha t A.$

13

Kinetic Theory of Gases

1. Kinetic Theory of Gases
Assumptions

(1) A gas consists of particles called molecules.

(2) The molecules are in random motion and obey Newton's laws of motion.

(3) The total number of molecules is large.

(4) The volume of the molecules is a negligible small fraction of the volume occupied by the gas.

(5) No appreciable forces act on the molecules except during a collision.

(6) Collisions are elastic and are of negligible duration.

2. Pressure of an Ideal Gas (P)

$$P = \frac{1}{3}\rho V^2$$

where ρ is the density of the gas V^2 is the moon square velocity of the molecules.

$$V = \sqrt{\frac{3P}{\rho}}$$

3. Kinetic Interpretation of Temperature

If M is molecular weight of the gas,

$$\frac{1}{2} MV^2 = \frac{3}{2} TR$$

The total translation kinetic energy per mole of the molecules of an ideal gas is proportional to temperature.

If m is the mass of a molecule

$$\frac{1}{2} mv^2 = \frac{3}{2} kT$$

$k = 1.380 \times 10^{-23}$ joule/molecule K, and is called the Boltzmann's constant.

The average translational kinetic energy of a molecule $\frac{3}{2} kT$.

4. Specific Heats of a Gas

$C_p - C_v = R$, where C_v is the molar heat capacity at constant pressure and C_v is the molar heat capacity at constant volume.

$$\frac{C_p}{C_v} = \gamma.$$

5. Equipartition of Energy

The available energy in a gas depends only on the temperature and distributes itself in *equal shares* to each of the independent ways in which the molecules can absorb energy. This theorem is called the theorem of equipartition of energy. Each such independent mode of energy absorption is called a *degree of freedom*.

Each degree of freedom of a molecule contributes an amount $\frac{1}{2}kT$ to the energy of the molecules.

A monatomic molecules the 3 degrees of freedom, a diatomic molecule has 5 degrees of freedom and a polyatomic molecule 6 degrees, unless it is linear. A linear polyatomic molecule has 5 degrees of freedom. In all these cases vibration of the atoms in the molecules is neglected.

C_p, C_v and γ may be calculated for different gases using the equipartition of energy.

Mean free path. The molecules in a gas are in random motion and they collide with one another. The mean free path L is the average distance between successive collisions.

$L = \dfrac{1}{\pi n d^2}$ where n is the number of molecules per unit volume, and d is the diameter of the molecule.

A more accurate, formula is

$$L = \frac{1}{\pi \sqrt{2} n d^2} = \frac{kT}{\pi \sqrt{2} p d^2}$$

6. Brownian Motion

The earliest and most direct experimental evidence for the reality of atoms was the proof of the atomic kinetic theory provided by the quantitative studies of Brownian Motion.

Brownian motion was discovered by Robert Brown in 1827. He found that pollen suspended in water shown a continuous random motion when viewed under a microscope. In 1905 Albert Einstein developed a theory of Brownian motion.

Brownian motions result from the impact of molecules of the fluid on the suspended particles, and these particles acquire the same mean kinetic energy as the molecules of the fluid.

7. Transport Phenomena

The coefficient of viscosity in a gas can be calculated from the kinetic theory. If m is the mass of a molecule, d its diameter, v the average speed.

$$\eta = \frac{mv}{3\sqrt{2\pi d^2}}$$

This equation enables us to calculate d from the experimental value of viscosity.

8. Maxwell's Velocity Distribution Law

The number dN_v of the molecules having speed lying between v and $v + dv$ is given by

$$dN_v = 4\pi a^3 N (e^{-E/kT}) v^2 dv$$

where N is the total number of molecules, and E is the kinetic energy of a molecule with a speed lying between the limits specified above.

$$E = \frac{1}{2} mv^2$$

Here, $\qquad a = \sqrt{\dfrac{m}{2\pi kT}}$

According to this distribution law

(i) the maximum velocity

$$v_m \sqrt{\frac{2kT}{m}}$$

(ii) the average velocity $\bar{v}$

$$= \sqrt{\frac{8kT}{m\pi}}$$

(iii) the root mean square speed v_{rms}

$$= \sqrt{\frac{3kT}{m}}$$

9. Van der Waal's Equation of State

Van der Waal allowed for the fact that the size of the molecule is not negligibly small. The actual volume available to a molecule is smaller. Thus, if the volume of the container is V, the actual volume available for the molecules is $(V - b)$. This is called free volume, and b is called Co-volume. He also allowed for the attractive forces between molecules by replacing p by $(p - a/V^2)$, where a is a constant.

The equation of state is

$$\left(p + \frac{a}{V^2}\right)(V - b) = RT$$

10. Critical Constructs

The critical Volume $V_c = 3b$

$$T_c = \frac{8a}{27bR}$$

$$P_c = \frac{a}{27b^2}$$

11. Liquefaction of Gases

For liquefying a gas its temperature must first be lowered to its critical temperature and below, and then compressed.

The chief problem in liquefaction is reaching the critical temperatures which are very low for the so-called permanent gases.

Cooling may be produced by a number of processes :

(i) *Boiling under reduced pressure* : A compressed vapour condenses and when it is allowed to evaporate under reduced pressure, it absorbs the latent heat from the tank.

(ii) *Cooling by adiabatic expansion of a gas* doing external work. When the gas expands adiabatically the work done is $\int pdv$, and this heat is taken from the gas, thus cooling it.

(iii) *Joule Kelvin Cooling* : Here the gas passes from high pressure to low pressure through a number of narrow orifices. If U_1 is the internal energy, V_1 the volume and P_1 the pressure on one side, and U_2, V_2, P_2 the corresponding values on the low pressure side

$$U_1 + P_1V_1 = U_2 + P_2V_2$$

The temperature change depends on the existence of intermolecular force and deviations from Boyle's Law. Cooling always takes place if the temperature initially is below a certain value called the temperature of inversion. Where calculates out to be $\dfrac{2a}{bR}$, a and b being Van der Waal's constants.

In actual process, a combination of these methods is used.

14

Calorimetry And Change of State

1. Basic Terms

Calorimetry. Unit of quantity of heat is expressed as a *calorie*. It is the quantity of heat required to raise the temperature of 1 gm of water through 1°C (more precisely from 16.5°C to 17.5°C).

Specific heat. It is the quantity of heat required to raise the temperature of 1 gm of the substance through 1°C. It is expressed as calorie per gm per deg C.

Thermal capacity. It is the quantity of heat required to raise the temperature of the body through 1°C

Thermal capacity = mass × specific heat × change in temp.

Heat gained or lost by a body. (No change of state takes place).

= mass × specific heat × change in temp.

Latent heat of fusion. It is the amount of heat required to melt one gm of substance (solid) without changing its temperature. *Example*—the latent heat of fusion of ice is 80 calories per gm.

Latent heat of vaporization. It is amount of heat required to vaporize 1 gm of the liquid without changing its temperature. *Example*—the latent heat vaporization of water is 536 calories per gm.

Specific heat of a gas at constant pressure. It is the amount of heat required to raise the temperature of 1 gm of the gas through 1°C at constant pressure. It is denoted by C_p.

Specific heat of a gas at constant volume. It is the amount of heat required to raise the temperature of 1 gm of the gas through 1°C at constant volume. It is denoted by C_v.

For a gm molecule $C_p - C_v = \dfrac{R}{J}$

where R is the universal gas constant and J the mechanical equivalent of heat.

Also $C_p/C_v = \gamma$ has the value 1.67 for monoatomic gases and 1.48 for diatomic gases.

2. Molar Specific Heats of Gases

Monoatomic gas. For a monoatomic gas, the molecules have only three independent degrees of freedom on account of only translatory motion. Therefore, for one mole of the gas at a temperature $T°k$, internal kinetic energy will be

$$U = \dfrac{3}{2} RT$$

So, $$C_v = \frac{dU}{dT} = \frac{3}{2}R$$

But, $$C_p = C_v + R$$

Therefor $$C_p = \frac{3}{2}R + R = \frac{5}{2}R$$

and so $$\gamma = \frac{C_p}{C_v} = \frac{5}{3} = 1.66.$$

Diatomic gas. For a diatomic gas, molecules have five degrees of freedom viz. three translatory and two rotatory ones. So, internal kinetic energy is

$$U = \frac{5}{2}RT$$

So $$C_v = \frac{d}{dT}U = \frac{5}{2}R$$

and $$C_p = C_v + R = \frac{5}{2}R + R = \frac{7}{2}R$$

Thus $$\gamma = \frac{C_p}{C_v} = \frac{7}{5} = 1.4.$$

In a like wise manner, one man calculate C_v, C_p and γ for higher atomed molecules of other gases knowing the degrees of freedom of the respective molecules.

The constant volume gas thermometre is the standard thermometre. The ideal gas scale is defined by the equation

$$T = 273.86\, P \lim_{tr \to 0} \frac{P}{P_{tr}}$$

where P is the pressure of the gas at temperature T, and P_{tr} is the pressure at the triple point.

This scale is identical with the (absolute thermodynamic) Kelvin Scale in the range in which the gas thermometer is used.

The International Practical Temperature Scale consists of a set of recipes for providing in practice the best possible approximation to the Kelvin Scale.

Molar heat capacities. The amount of heat per mole necessary to increase the unit temperature is the molar heat capacity.

If ΔQ is the heat required to produce a rise in temperature ΔT in m gram of a gas at constant volume, then

$$C_v = \frac{\Delta Q}{\mu \Delta T} \text{ constant volume}$$

and similarly, $C_p = \dfrac{\Delta Q}{\mu \Delta T}$ constant pressure

where $\mu = \dfrac{m}{M}$ number of mole

(M = molecular weight)

The ratio of C_p/C_v is denoted by γ. The value of γ for monoatomic gases is 1.67 and nearly 1.40 for diatomic gases. The difference of C_p and C_v is given by

$$C_p - C_v = R$$

This is known as Mayer's relation.

15

Thermodynamics

1. Equivalence of Work and Heat

Joule's experiments establish a relation between heat and work. According to it 4.18×10^3 joules of work (performed in different manners in different substances) produces the same temperature rise as 1 kilocalorie (amount of heat required to raise the temperature of 1 kg of water through 1°C) of heat. Thus Joule concluded that 4.18×10^3 joules of work is equivalent to 1 kilocalorie of heat

$$1 \text{ kilo-calorie} = 4.18 \times 10^3 \text{ joule}$$

or
$$1 \text{ calorie} = 4.18 \text{ joule}$$

2. First Law of Thermodynamics

Let a system change from state i, to state f in definite way, the heat absorbed by the system being Q and the work done by the system being W. It is found that the quantity $(Q - W)$ depends only on the initial and final coordinates and not at all on the path between the points. Hence there is a function of the thermodynamic coordinates whose final value minus its initial value equals the change $Q - W$ in the process. We call this function, the internal energy function U.

Thus $\quad U_f - U_i = \Delta U = Q + W.$

This is called the first law of thermodynamics. If the system undergoes only infinitesimal change, the law may be written.

$$dU = dQ - dW$$

3. Concept of Internal Energy

Internal energy of a system refers to the energy possessed by the system due to (i) molecular motion, and (ii) molecular configuration.

The energy due to molecular motion is known as internal kinetic energy and is expressed by U_k. The energy due to molecular configuration is known as internal potential energy and is expressed as U_p. Hence

$$\Delta U = \Delta U_k + \Delta U_p$$

If intermolecular forces are absent

$$\Delta U_p = 0$$

∴ $\quad \Delta U = \Delta U_k$

If m is the mass of the system, C_v specific heat at constant volume and ΔT is change in temperature, then

$$\Delta U = m\, C_v\, \Delta T$$

In terms of molar heat capacity C_v, the internal energy can be expressed as

$$\Delta U = M\, C_v\, \Delta T$$

where $\quad M = $ molecular weight.

For μ-moles of an ideal gas

$$\Delta U = \mu\, C_v\, \Delta T = \frac{m}{M} C_v\, \Delta T$$

4. Concept of Work

The workdone is defined as the flow of energy without a difference of temperature.

The workdone by the gas in expanding from volume V_1 to V_2 will be given by the sum of all terms of i.e.,

$$W = \sum_{v_1}^{V_2} (P \times V)$$

If during expansion of the gas, the pressure P is kept constant, then $P - V$ graph will be straight line (see Fig. 1a) parallel to volume axis. The workdone in this will be given by

$$W = P \sum_{V_1}^{V_2} \Delta V = P(V_2 - V_1)$$

= area enclosed between AB and volume axis

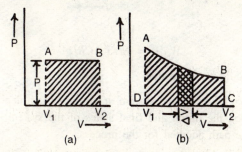

Fig. 1.

If the pressure of the gas is continuously changing the $P-V$ graph will be a curve as shown in Fig. (1b). In this case, the workdone will be given by

$$W = P \sum_{V_1}^{V_2} P \times \Delta V = \text{Area } ABCD$$

Thus, the workdone can be obtained directly by the area enclosed between the $P-V$ curve and volume axis.

When a system after passing through different states returns to its original state, then it is known as a cyclic process. The cyclic process is shown in Fig.(2). In case of cyclic process, the net workdone on the system or by the system is equal to the area enclosed by the curve. Thus

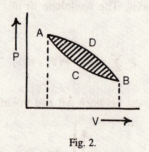

Fig. 2.

$$W_{\text{cyclic}} = \text{Area } ACBDA$$

It should be noted that

(i) Workdone by the system or on the system depends not only upon the initial and final states of the system but also upon the path adopted for the process.

(ii) If in a process, the volume of the system increases then the work is done by the system. The workdone by the system is taken as positive. On the other hand, if the volume of the system decreases then the work is done on the system. The workdone on the system is taken as negative.

5. Second Law of Thermodynamics

Clausius : It is impossible for any cyclic machine to produce no other effect than to convey heat continuously from one body to another at a higher temperature.

Kelvin and Planck. 'A transformation whose only final result is to transform into work heat extracted from a source which is at the same temperature throughout is impossible.'

OR

It is impossible to get a continuous supply of work by cooling a body to a temperature lower than that of the coldest of the surroundings.

Reversible process. A versible process is one that, by a differential change in the environment, can be made to retrace its path.

An isothermal operation can be approached if the process takes place extremely slowly so that heat may be absorbed from the surroundings during expansion and rejected to the surroundings during compression, at such a rate that the temperature remains practically constant.

An adiabatic operation may be approached in practice if it takes place so quickly that no time is allowed for the heat to enter or leave the gas.

A *Cycle* consists of a series of operations, taking place in a certain order, which restore initial conditions.

If the cycle of operations is plotted in a $P - V$ diagram, the net work done by the working substance during one cycle is given by the enclosed area of the diagram.

6. Types of Cycle

(i) Carnot cycle—a constant temperature cycle.

(ii) Otto cycle—a constant volume cycle.

(iii) Diesel cycle—a constant pressure cycle.

Efficiency of Carnot engine

If Q_1 is the heat absorbed at temperature T_1 from the source, and Q_2 the heat rejected at temperature T_2 to the sink, the efficiency of the engine is

$$= \frac{Q_1 - Q_2}{Q_1} = \frac{T_1 - T_2}{T_1}$$

Carnot's theorem. The efficiency of a thermal engine operation according to a reversible Carnot cycle is independent of the working substance and depends only on the two operating temperatures. Further efficiency of a reversible engine is more than the efficiency of all other engine working in the same temperature range.

Efficiency of Otto cycle

$$\eta = 1 - \left(\frac{1}{\rho}\right)^{\gamma - 1}$$

where ρ is the compression ratio, γ the ratio of specific heats of gas.

Efficiency of Diesel cycle

$$\left(1 - \frac{1}{\rho}\right)^{\gamma - 1} \frac{(K\gamma - 1)}{\gamma(K - 1)}$$

Definitions and Formulae in Physics

where $$K = \frac{\text{adiabatic compression ratio}}{\text{adiabatic expansion ratio}}$$

For a given compression ratio, the Otto Cycle efficiency is greater than the Diesel Cycle efficiency.

However, in Diesel engines, compressions is not limited by the ignition temperature of the fuel. Hence, in practice a greater operating efficiency can be attained in Diesel engines which are designed for higher compression ratios.

7. Entropy

The difference in entropy between two states A and B is the value of the integral

$$\int_A^B \frac{\partial Q}{T}$$

taken along any reversible path connecting the two states. For and infinitesimal change

$$\frac{\partial Q}{T} = dS$$

where dS is the change in entropy.

Entropy increases if dQ denotes absorption of heat. Entropy decreases if dQ denotes rejections of heat. Adiabatic reversible transformations/are at constant entropy.

8. Maxwell's Relations

$$\left(\frac{\partial S}{\partial V}\right)_T = \left(\frac{\partial P}{\partial T}\right)_V$$

$$\left(\frac{\partial S}{\partial P}\right)_T = \left(\frac{\partial V}{\partial T}\right)_P$$

$$\left(\frac{\partial T}{\partial V}\right)_S = \left(\frac{\partial S}{\partial T}\right)_V$$

$$\left(\frac{\partial T}{\partial P}\right)_S = \left(\frac{\partial V}{\partial S}\right)_P$$

These equations do not refer to any process but hold at any equilibrium state.

9. Refrigerator

Refrigerator is a heat engine running in backwards direction. In this case, the working substance (a gas) takes heat from a cold body (inner space of the refrigerator) and gives out to a hotter body (external atmosphere). For this purpose,

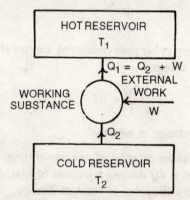

external energy (electrical energy) is required. No refrigerator has yet been designed which can transfer heat from a cold body to a hot body without using an external energy source. Thus, it is impossible for a self-acting machine, unaided by any external agency to transfer heat from a cold body to a hot body.

The coefficient of performance of a refrigerator

$$\beta = \frac{\text{heat extracted from cold reservoir}}{\text{workdone on refrigerator}}$$

$$= \frac{Q_2}{W} = \frac{Q_2}{Q_1 - Q_2} = \frac{1}{(Q_1/Q_2) - 1}$$

16

Isothermal and Adiabatic Changes

1. Isothermal Change

When a system undergoes a physical change under the condition that temperature of the system remains constant then such a change is called as isothermal change. Under such a change,

$$PV = \text{constant}$$

2. Adiabatic Change

It is that change which occurs within a system in such a way that no heat can enter or leave the system.

For a perfect gas in adiabatic condition, the following equation holds

$$PV^\gamma = \text{constant}$$

where γ is the ratio of the specific heat of the gas at constant pressure and the specific heat at constant volume, ($\gamma = C_p/C_v$).

3. Isothermal and Adiabatic Curves

The graph drawn between the pressure and the volume of a given mass of a gas undergoing an isothermal change is called isothermal curve; and that for a gas undergoing an adiabatic change is called adiabatic curve.

For an ideal gas, the ratio of the slopes of the adiabatic and isothermal curves is equal to the ratio γ, i.e.,

Slope of adiabatic curve = $\gamma \times$ slope of isothermal curve.

Since the value of γ is always greater than one, hence the adiabatic curve is more stepper as compared to isothermal curve.

4. Work Done Under Isothermal Change

The workdone dW under isothermal change is given by

$$dW = RT \times 2.303 \, \text{Log}_{10} (V_f/V_i)$$

Where V_i and V_f are the initial and final volumes of the gas under isothermal change.

5. Work Done Under Adiabatic Change

The workdone dW under adiabatic change is given by

$$dW = \frac{R(T_2 - T_1)}{(\gamma - 1)} = \frac{P_2 V_2 - P_1 V_1}{(\gamma - 1)}$$

where T_1 and T_2 are initial and final temperatures.

17

Transmission of Heat and Thermal Conduction

1. Processes of Transmission of Heat
(i) Conduction, (ii) Convection, (iii) Radiation.

(i) **Conduction.** Conduction is the flow of heat through an unequally heated body from places of higher temperature to those of lower temperature without actual transference of particles constituting the body. Almost all solids are good conductors of heat.

(ii) **Convection.** It is the process by which heat is transferred from one place to another in a medium by the movement of particles of the medium. Convection occurs in fluids (liquid and gases).

(iii) **Radiation.** It is the process by which heat is transferred from one place to another without the agency of any intervening medium. Heat from the sun comes on the earth by radiation.

2. Steady State

When the temperature of various points of the bar is changing, the state is said as variable state. After sometime a state is reached when the temperature of each cross-section becomes steady. This state is known as steady state. In this state, any heat received by any cross section is partly conducted to the next section and partly radiated i.e., no heat is absorbed by the cross section.

3. Coefficient of Thermal Conductivity

Consider a rectangular slab of which two opposite faces have each an area A and they are at a distance d from each other. If one face is at a higher temperature T_1 and the opposite face at a lower temperature T_2, then the quantity of heat Q passing from the hotter face to the colder face is given, by the following :

$$Q = KA \frac{T_1 - T_2}{d} t$$

where t is the time and K is a constant depending upon the material of the slab.

This constant K is called the *coefficient of thermal conductivity* and the quantity $\frac{T_1 - T_2}{d}$ is called the temperature gradient.

So the *coefficient of thermal conductivity* is the amount of heat transmitted per second per unit area per unit temperature gradient.

Thermal resistance. The thermal resistance of a body is a measure of its opposition to the flow of heat through it. It is defined as

$$\text{Thermal resistance} = \frac{\text{temperature difference at the two ends}}{\text{rate of flow of heat through it}}$$

It can easily be shown that

$$\text{Thermal resistance} = \frac{\text{length or thickness of the material}}{\text{thermal conductivity} \times \text{area}}$$

$$= \frac{1}{KA}.$$

4. Newton's Law of Cooling

According to this law, the rate of loss of heat of a body is directly proportional to the temperature difference between the body and the surroundings, i.e.

$$\frac{dQ}{dt} = -k(T - T_0)$$

where T = temperature of body and T_0 = temperature of surroundings.

This law is true for small temperature difference only.

Radiation. It is process of transfer of heat in which no material medium is required.

Black body. It is that which absorbs all the radiant energy falling on it.

Definitions and Formulae in Physics

Good radiator. Is a *good absorber,* and a *bad radiator* is a *bad absorber.*

5. Stefan's Law of Radiation

It states that the total radiant energy radiated per sec by a black body is proportional to the fourth power of its absolute temperature.

$$E = \sigma a T^4$$

- E = energy radiated
- T = absolute temperature
- σ = Stefan's constant
- a = area of surface

Stefans' Constant

$$\sigma = 5.67 \text{ joules/m}^2 - \sec (^\circ K)^4.$$

Emissive power. The emissive power of a body, at a given temperature and for a particular wavelength is defined as the radiant energy emitted per second by unit surface area of the body per unit wavelength range.

Absorptive power. The absorptive power of a body, at a given temperature and for a particular wavelength is defined as the radiant energy absorbed per second by unit surface area of the body to the total energy falling per unit time on the same area.

6. Kirchhoff's Law

This law states that the ratio of the emissive power to the absorptive power for radiation of a given to wavelength is the same for all bodies at the same temperature and is equal to the emissive power of a perfectly black body at that temperature.

7. Wien's Displacement Law

The wavelength corresponding to maximum energy is an inversely proportional to the Kelvin temperature and is given by

$$\lambda_m \propto \frac{1}{T}$$

or $\qquad \lambda_m T = \text{constant}$

The value of this constant is found to be 2.93×10^{-3} mK.

Solar constant. Solar constant is defined as the amount of energy received from the sun by the earth per minute per cm^2 of surface placed normally to the sun's rays at mean distance of the earth from the sun in the absence of atmosphere. The value of Solar constant is 1.937 cal. cm^{-2} min^{-1}.

18

Simple Harmonic Motion

1. Types of Motion

Periodic motion. A motion that repeats itself after equal intervals of time is known as periodic motion.

Oscillatory motion. A body is said to possess oscillatory or vibratory motion if it moves back and forth repeatedly about a mean position.

Simple harmonic motion. This is a special type of periodic motion where in the body moves again and again over the same path about a fixed point (equilibrium position) in such a way that it is acted upon by a restoring force proportional to its displacement from the mean position.

2. Characteristics of Simple Harmonic Motion

(i) The motion is periodic.

(ii) The motion is along a straight line about the mean or equilibrium position.

(iii) The acceleration is proportional to displacement.

(iv) Acceleration is directed towards the mean or equilibrium position.

3. Simple Pendulum

A simple pendulum is made up of heavy point mass suspended by a weightless inextensible and perfectly flexible string from a rigid support. Such an ideal pendulum is not possible in practice but a heavy job suspended by a light inextensible thread works as simple pendulum.

The time period T of simple pendulum is given by

$$T = 2\pi\sqrt{\left(\frac{l}{g}\right)}$$

where l is the length of simple pendulum.

When the time period of a simple pendulum is 2 second, it is known as a second pendulum.

When one end of a weightless spring is attached to a mass m and other end is fixed, the periodic time T of vibrating system is given by

$$T = 2\pi\sqrt{\left(\frac{m}{k}\right)}$$

where k is the force constant of the spring.

Geometrical Representation of S.H.M. Simple harmonic motion can be easily represented as the projection on any diameter of a point moving in a circle with uniform speed.

Displacement. The displacement y is given by

$$y = a \sin(\omega t \pm \phi)$$

Definitions and Formulae in Physics

where a is the amplitude, ω is the angular velocity and ϕ is the initial phase.

If the initial phase is zero, then

$$y = a \sin \omega t.$$

The phase of a vibrating particle at any particular instant refers to its state as regards its position and its direction of motion at the instant. The phase difference between two vibrations indicates how much the two vibrations are out of step with each other.

Velocity. The velocity of a particle executing S.H.M. is the rate of change of displacement, and is given by.

$$v = \frac{dy}{dt} = \omega \sqrt{(a^2 - y^2)}$$

Acceleration. The acceleration of a particle executing S.H.M. is the rate of change of velocity and is given by

$$\text{acceleration} = \frac{dv}{dt} = -\omega^2 y.$$

Periodic time. It is the time taken to complete the simple harmonic motion once, it is given by

$$T = \frac{2\pi}{\omega} = 2\pi \sqrt{\left(\frac{\text{displacement}}{\text{acceleration}}\right)}$$

4. Energy of Simple Harmonic Motion

When a particle performs simple harmonic motion, its energy changes between kinetic energy and potential energy such that the sum of two remains constant. At any particular instant,

Kinetic energy $= \frac{1}{2} mv^2 = \frac{1}{2} m\omega^2 (a^2 - y^2)$

Potential energy $= \frac{1}{2} ky^2 = \frac{1}{2} m\omega^2 y^2$

Total energy = Kinetic energy + Potential energy

$$= \frac{1}{2} m\omega^2 (a^2 - y^2) + \frac{1}{2} m\omega^2 y^2$$

$$= \frac{1}{2} m\omega^2 a^2$$

where $\qquad \omega = 2\pi n \qquad$ (n = frequency)

5. Motion of a Body Suspended by a Spring

Consider a light spring (force constant K) is hanged from a rigid support A as shown in Fig. (1). A mass m is suspended from the lower end. If the mass is pulled down and released

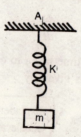

Fig. 1.

then it executes S.H.M. The time period of the system is given by

$$T = 2\pi \sqrt{\frac{m}{K}} \qquad \ldots(1)$$

The following special cases may arise.

(i) if m_s be the mass of the spring, then the expression of time period given by eq. (1) is modified as

$$T = 2\pi \sqrt{\frac{m + (m_2/3)}{K}} \qquad ...(2)$$

(ii) If a spring of force constant K is divided into n equal parts and one such part is attached to a mass m, then the time period is given by

$$T = 2\pi \sqrt{\frac{m}{nk}} \qquad ...(3)$$

(iii) If two springs of force constant K_1 and K_2 are connected in parallel and a mass m is attached to them, then the time period is given by

$$T = 2\pi \sqrt{\frac{m}{K_1 + K_2}} \qquad ...(4)$$

where $K = K_1 + K_2$

(iv) If two springs of force constants K_1 and K_2 are connected in series and a mass m is attached to them, then the time period is given by

$$T = 2\pi \sqrt{\frac{m(K_1 + K_2)}{K_1 K_2}} \qquad ...(5)$$

where $\dfrac{1}{K} = \dfrac{1}{K_1} + \dfrac{1}{K_2} = \dfrac{K_2 + K_1}{K_1 K_2}$

or $K = \left(\dfrac{K_1 K_2}{K_1 + K_2}\right)$

(v) If two masses m_1 are connected by a spring then the time period is given by

$$T = 2\pi \sqrt{\left(\frac{\mu}{K}\right)}$$

where $$\mu = \frac{m_1 m_2}{m_1 + m_2}$$

Here μ is known as reduced mass.

19

Wave Motion and Speed of Mechanical Waves

1. Importance of Simple Harmonic Motion

Sound is created by movement of bodies. Therefore, to study sound one is required to study movement of bodies. Movement are innumerous and so apparently this study looks to be an enormous task. But, thanks to Simple Harmonic Motion, the situation is not so bad as it looks. This is so because SHM is such a basic motion that it is possible to break up any motion into a sum of a large number of SHMS (reference, Fourier expansion of functions). Thus, just the study of SHM is sufficient and so the importance of SHM.

2. Simple Harmonic Motion

Any system of mass m upon which a force $F = -kx$ where k is a constant of the system) acts will be governed by the following equation :

$$\frac{d^2x}{dt^2} + \frac{k}{m}x = 0$$

k is called the force constant, and is the restoring force per unit displacement.

The solution to this equation is

$$x = A \cos (\omega t + \delta)$$

where $$\omega^2 = \frac{k}{m}.$$

The period $$T = \frac{2\pi}{\omega} = 2\pi \sqrt{\frac{m}{k}}.$$

The quantity ω is called the *angular frequency*. A is the amplitude (the maximum displacement). The quantity $(\omega t + \delta)$ is called the phase.

The velocity of the particle.

$$v = \frac{dx}{dt} = -\omega A \sin (\omega t + \delta)$$

The acceleration

$$v = \frac{d^2x}{dt^2} = -\omega^2 A \cos (\omega t + \delta)$$

$$= -\omega^2 x$$

We note that

$$v^2 = \omega^2(A^2 - x^2)$$

The maximum value of v occurs when $x = 0$

$$v_m = \pm \omega A$$

The maximum value of acceleration is when $x = A$

$$a_m = \omega^2 A$$

At any instant, the kinetic energy

$$= \frac{1}{2} mv^2 = \frac{1}{2} m\omega^2 (A^2 - x^2)$$

At any instant, the potential energy

$$= \frac{1}{2} kx^2$$

The total energy

= the maximum kinetic energy

$$= \frac{1}{2} m\omega^2 A^2$$

It is also equal to the maximum potential energy

$$= \frac{1}{2} kA^2$$

(i) **Wave motion.** A wave motion is defined as a disturbance which is handed over from one part of the medium to the next due to the repeated motion of the medium particles about their mean positions.

(ii) **Longitudinal waves.** Longitudinal wave is a wave motion in which the particles of the medium vibrate along the direction of propagation of the wave. The most common example of a longitudinal wave is a sound wave.

(iii) **Transverse waves.** Transverse wave is a wave motion in which the particles of the medium vibrate at right angles to the direction of propagation of wave. Waves of pluked strings, electromagnetic waves, light waves are the examples of transverse waves.

(iv) **Frequency (n) or (v).** The number of vibrations performed by a particle in one second is defined as frequency.

(v) **Wavelength (λ).** The distance travelled by the wave in one time period i.e., T seconds is called wavelength. The wavelength of the wave is shown in Fig.

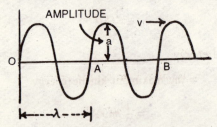

(vi) **Time period (T).** This is defined as the time required to complete one vibration.

(vii) **Amplitude (A).** The maximum displacement of the particle on either side of its mean position is called its amplitude.

(viii) **Velocity (v).** The distance travelled by the wave in one second is defined as velocity.

(ix) **Relationship between velocity, frequency and wavelength.** The relationship between velocity v, frequency n and wavelength λ is expressed as

$$v = n\lambda$$

Here it should be remembered that

$$\text{Frequency} = \frac{1}{\text{Time period }(T)}$$

(x) **Phase difference.** The phase difference between two waves is defined as how much the difference between the two

Definitions and Formulae in Physics

waves is defined as how much the two waves are out of step with each other or by how much time one wave is ahead of the other wave.

$$\text{Phase difference} = \left(\frac{2\pi}{\lambda}\right) \times \text{path difference}.$$

Phase Relations Amongst Displacement, Velocity and Acceleration for a body in SHM

The time dependence of displacement is given as

$$y = A \cos(\omega t + \delta)$$

Differentiating this once and twice with respect to time, one gets the velocity and acceleration as

$$v = \frac{dy}{dt} = -A\omega \sin(\omega t + \delta)$$

$$= A\omega \cos\left(\omega t - \frac{\pi}{2} + \delta\right)$$

$$a = \frac{d^2y}{dt^2} = -A\omega^2 \cos(\omega t + \delta)$$

$$= A\omega^2 \cos(\omega t - \pi + \delta)$$

We note that phases of displacement velocity and acceleration at time t are respectively ωt, $\left(\omega t - \frac{\pi}{2}\right)$ and $(\omega t - \pi)$. It is therefore clear that phase difference between displacement and velocity is $\frac{\pi}{2}$, that between velocity and acceleration is $\frac{\pi}{2}$ and further that between displacement and acceleration is π. Obviously, when displacement is minimum, velocity is maximum and *vice versa*. However, when displacement is minimum, acceleration is also minimum and when displacement is maximum, acceleration is maximum but for that they are directed opposite, being of opposite signs.

Alternatively, one can write the equation of simple Harmonic Motion as

$$y = A \sin(\omega t + \delta)$$

also, under these circumstances,

$$v = \frac{dy}{dt} = A\omega \cos(\omega t + \delta)$$

$$= -A\omega \sin\left(\omega t + \delta - \frac{\pi}{2}\right)$$

and

$$a = \frac{d^2y}{dt^2} = -A\omega^2 \cos(\omega t + \delta)$$

$$= A\omega^2 \sin(\omega t + \delta - \pi)$$

3. Wave Motion

The equation $y = f(x - vt)$ represents a pulse travelling in the positive direction of x with velocity v.

The equation $y = f(x + vt)$ represents a pulse travelling in the negative x direction. If the wave is sinusoidal, the equation is

$$y = y_m \sin 2\pi \left(\frac{x}{\lambda} - \frac{t}{T}\right)$$

This is a progressive sinusoidal wave. Here T is the period of vibration, and λ is the distance travelled by the wave in time t.

$$\text{Velocity } v = \frac{\lambda}{T} = f\lambda,$$

where f is the frequency.

If we write

$$= \frac{2\pi}{\lambda} \text{ and } \omega = \frac{2\pi}{T}$$

$y = y_m \sin(kx - \omega t)$ for a sine wave travelling in the positive x direction, and

$y = y_m \sin(kx + \omega t)$ for a wave travelling in the negative x direction.

In times of ω and k, $v = \omega k$.

The wave equation may also be written as

$$y = y_m \sin \omega(x/v - t)$$
$$y = y_m \sin k(x - vt)$$

In the above expression it has been assumed that at the point $x = 0$, $y = 0$ at time $t = 0$. This need not be. The general case is

$$Y = y_m \sin(kx - \omega t - \phi)$$

Where ϕ is called the phase constant.

There are two types of waves viz., the longitudinal and transverse waves. In the case of longitude waves particles of the medium vibrate in the direction of propagation of the wave itself whereas in the case of transverse waves, they vibrate in a direction perpendicular to the direction of propagation. While the fist category involves formation of condensations and rarfactions, the second category involves formation of crests and troughs. Examples of longitudinal waves are sound waves, longitudinal waves in metal rods etc., where as examples of transverse waves are ripples on water surface, vibrations in strings etc.

4. Mechanical Waves

It is a disturbance which travels in the material medium without changing its shape with a definite velocity. The particles of the medium vibrate in a direction perpendicular to the direction of wave propagation. These waves transfer energy and momentum through the limited motion of the particles with the medium remaining at its own place. These waves require a medium. Longitudinal and transverse waves are the examples of mechanical waves.

5. Ultrasonic Waves

The human ear can hear the sound waves between 20Hz to 203Khz. This range is known as audible range. The sound waves having frequencies above the audible range are known as *ultrasonic waves or supersonic waves*. The wavelengths of ultrasonic waves are very small as compared to audible sound. The wavelength of ultrasonic waves in air will be 1.8 cms at 20Hz and becomes much less at higher frequencies (0.035 m at 1 Mhz). These waves travel with the velocity of sound waves. The sound waves which have frequencies less than the audible range are called *infrasonic waves*.

6. Equation of a Plane Progressive Wave

The displacement y of a particle of a medium at any time t in a wave motion at a distance x from the origin is expressed as

$$y = a \sin \frac{2\pi}{\lambda} (vt - x) \qquad \ldots(1)$$

This is the equation of progressive wave propagating along x-axis.

Equation (1) can also be expressed as

$$y = a \sin 2\pi \left(\frac{t}{T} - \frac{x}{\lambda} \right) \qquad \ldots(2)$$

$$y = a \sin 2\pi n \left(t - \frac{x}{v} \right) = a \sin \omega \left(t - \frac{x}{v} \right) \qquad \ldots(3)$$

and $\quad y = a \sin(\omega t - kx)$, where $k = \dfrac{\omega}{v} = \dfrac{2\pi}{\lambda} \qquad \ldots(4)$

k is known as propagation constant.

7. Velocity of Longitudinal Waves in Elastic Medium

(i) Newton's formula. On the basis of theoretical investigations, Newton proved that when longitudinal wave moves in an elastic medium, the velocity v is given by

$$v = \sqrt{\left(\frac{E}{d}\right)}$$

where E is the modulus of elasticity of the medium and d is its density.

(ii) For solid medium. For a solid medium, the modulus of elasticity E is equal to Y, the Young's modulus of elasticity. Hence, the velocity of longitudinal wave in a solid medium is given by

$$v = \sqrt{\left(\frac{Y}{d}\right)}.$$

For example, in case of iron, $Y = 2.0 \times 10^{11}$ newton/metre^2 and $d = 7.7 \times 10^3$ kg/m^3. Hence

$$v = \sqrt{\left(\frac{2.0 \times 10^{11}}{7.7 \times 10^3}\right)} = 5095 \text{ m/sec.}$$

(iii) **For liquid medium.** For a liquid medium, the modulus of elasticity E is equal to B, the bulk modulus of liquid. Hence

$$v = \sqrt{\left(\frac{B}{d}\right)}.$$

For example, in case of water, $B = 2 \times 10^9$ newton/metre^2 and $d = 1 \times 10^3$ kg/metre^3. Hence

$$v = \sqrt{\left(\frac{2 \times 10^9}{1 \times 10^3}\right)} = 1414 \text{ m/sec.}$$

(iv) **For gaseous medium.** For a gaseous medium. Newton assumed that the propagation of longitudinal wave is an isothermal process i.e., temperature remains constant. The modulus of elasticity E is equal to the pressure of the gas. Hence

$$v = \sqrt{\left(\frac{P}{d}\right)}$$

For example, in case of air, at N.T.P., $P = 1.013 \times 10^5$ newton/metre^2 and $d = 1.293$ kg/m^3.

$\therefore$ Velocity of sound at 0°C = $\sqrt{\left(\dfrac{1.013 \times 10^5}{1.293}\right)}$

$$= 280 \text{ m/sec.}$$

Experimentally it is observed that 0°C the velocity of sound is 332 metre/sec. Obviously, Newton's formula is not correct.

(v) **Laplace's correction.** Laplace pointed out that the propagation of sound in gaseous medium is not isothermal process but it is an adiabatic process. Under adiabatic change the modulus of elasticity of gas $E = \gamma P$, where γ is the ratio of specific heat of the gas at constant pressure to that at constant volume. Hence,

$$v = \sqrt{\left(\frac{\gamma P}{d}\right)}$$

$$= \sqrt{\left(\frac{1.41 \times 1.013 \times 10^5}{1.293}\right)} \quad (\because \gamma = 1.41 \text{ for air})$$

$$= 332.5 \text{ m/sec.}$$

This shows that Laplace's correction is true.

It is quite obvious that the velocity of longitudinal waves in

Solid medium > liquid medium > gaseous medium.

4. Effect of Temperature, Pressure and Humidity on Velocity of Sound

(1) **Effect of temperature.** If v_t and v_0 be the velocities of sound at $t°C$ and $0°C$ respectively, then

$$\frac{v_t}{v_0} = \sqrt{\left(\frac{T_t}{T_0}\right)}.$$

where T_t and T_0 are absolute temperatures.

Thus the velocity of sound is directly proportional to the square root of the absolute temperature.

Again $v_t = v_0(1 + \alpha t)$ where $\alpha = \dfrac{1}{273}$ for all gases and is known as coefficient of volume expansion.

(2) **Effect of pressure.** *If the temperature of the gas remains constant, the velocity of sound does not change with a change of pressure because P/d remains constant.*

(3) **Effect of humidity.** If v_m and v_d be the velocities of sound in moist air and in dry air respectively, then

$$v_m > v_d$$

Thus the velocity of sound in moist air is greater than in dry air.

Under the same conditions of temperature and pressure the velocity of sound in gases **varies inversely as the square root of their densities.**

$$\frac{v_1}{v_2} = \sqrt{\frac{d_2}{d_1}}$$

For example, $\dfrac{\text{velocity in hydrogen}}{\text{velocity in oxygen}}$

$$= \sqrt{\frac{d_{oxygen}}{d_{hydrogen}}} = \sqrt{\frac{16}{1}} = \frac{4}{1}.$$

The speed of sound in air increases approximately by 0.6 metre per second for 1°C rise in temperature.

5. Conditions of Interference

The following conditions must be fulfilled for interference of two waves.

(i) The sources must have the same frequency for interference to be sustained.

(ii) The amplitude of waves must be equal.

(iii) The displacement produced by two waves should be along the same lines at a point.

(iv) For destructive interference phase difference between two waves should be odd multiple of $\lambda/2$.

6. Beats

When two sound sources of slightly different frequencies and of moderate intensities are sounded simultaneously, the intensities of the resultant sound periodically fluctuates, thus causing waxing and waving. *The Phenomenon of waxing and waving in the loudness of sound is called beats.*

7. Stationary Waves

When two identical progressive waves travel through a medium along the same line in opposite direction, they superimpose to produce new type of waves which appear stationary in space. These waves are called stationary waves.

8. Resonance

It is a special case of forced vibrations. When the natural frequency of a vibrating body is equal to the frequency of the periodic force then the body vibrates with maximum amplitude. This is the phenomenon of resonance.

9. Loudness

Loudness is that characteristics of sound which determines the degree of sensation produced in the ear.

20

Vibrations of Strings and Air Columns

1. Simple Vibrating System

Transverse Vibrations of a string

$$f_1 = \frac{1}{2l} \sqrt{\frac{T}{m}}$$

where f_1 is the fundamental.

$$= \frac{1}{2l} \sqrt{\frac{T}{\pi r^2 d}}$$

where r is the radius of the string, and d is the density.

If the string vibrates in x loops,

$$= \frac{x}{2l} \sqrt{\frac{T}{m}}$$

where f_x is the xth harmonics.

(i) **Vibrations in strings.** When a stretched string is plucked, transverse vibrations take place.

Formula : $$n = \frac{1}{2l}\sqrt{\frac{T}{m}}$$

where n = frequencies of the fundamental note.
 l = length of the string.
 T = tension of the string.
 m = mass of the string per unit length.

2. Laws of Vibrations in Strings

The above formula gives the following laws

(i) The frequency of vibration is inversely proportional to the length of the string

$$n \propto 1/l.$$

(ii) The frequency is directly proportional to the square root of the tension

$$n \propto \sqrt{T}.$$

(iii) The frequency is inversely proportional to the square root of the mass per unit length of the string

$$n \propto 1/\sqrt{m}.$$

The velocity of transverse wave in the string is equal to

$$\sqrt{T/m}.$$

The wavelength (fundamental) = $2l$.

(ii) **Harmonics.** The term *harmonics* refers to a series of frequencies which have whole number ratios. If n is the lowest *fundamental* frequency called the first harmonic, then the second harmonic has frequency $2n$, third $3n$, fourth $4n$ and so on.

The harmonics and overtones produced in the stretched string are shown in Fig. (1).

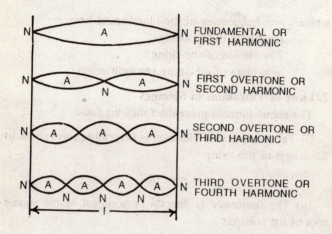

When the string is plucked in the middle, it vibrates with nodes at the ends and antinode in the middle. The tone emitted in this case is known as fundamental or first harmonic. The frequency n_1 is given by

$$\therefore \quad n_1 = \frac{1}{2l} \sqrt{\left(\frac{T}{m}\right)} \qquad \ldots(1)$$

If the wire is plucked at one fourth of its length, the string vibrates in two segments. The frequency of vibrations of string n_2 is given by

$$n_2 = \frac{1}{l} \sqrt{\left(\frac{T}{m}\right)} = 2n_1 \qquad \ldots(2)$$

when the wire vibrates in three segments, then

Definitions and Formulae in Physics

$$n_3 = \frac{3}{2l} \sqrt{\left(\frac{T}{m}\right)} \; 3n_1 \qquad \ldots(3)$$

Similarly when string vibrates in four segments,

$$n_4 = \frac{4}{2l} \sqrt{\left(\frac{T}{m}\right)} = 4n_1$$

$$n_1 : n_2 : n_3 : n_4 \ldots = 1 : 2 : 3 : 4 \ldots$$

Here it should be remembered that $n_p = pn_1$ where p is number of loops of segments.

(iii) **Overtones.** The harmonics other than the fundamental are called *overtones*. Then first overtone is the second harmonic and the second overtone is the third harmonic and so on.

A stretched string is capable of having all the harmonis, even as well as a odd.

(iv) **Vibrations in air columns.** (Resonance tubes)

(v) **Closed tubes.** Here one end of the tube is closed and the other open. If l is the length of the air column,

then $\qquad l = \lambda/4$ or $\lambda = 4l$.

So the lowest (fundamental) frequency

$$n_1 = \frac{V}{4l} \qquad \text{(where } V \text{ is the velocity of sound in air).}$$

The closed tube contains only odd harmonics, namely $3n_1$, $5n_1$, etc.

(vi) **Open tubes.** Here both ends of the tube are open if l is the length of the tube, then the fundamental mode of vibrations has

$$\lambda = 2l.$$

So the fundamental frequency $n_1 = \dfrac{V}{2l}$

In such a tube all the harmonics are possible —, even and odd. Thus, we have frequencies of harmonics as $2n_1$, $3n_1$, $4n_1$ etc., in the open tube.

(vii) **Resonance.** If sound waves fall on a body and the frequency of the wave is equal to the natural frequency of vibration of the body, then the body also begins to vibrate with a large amplitude. This phenomenon is called *resonance*.

(viii) **End correction.** In the above calculations it was assumed that the antinodes coincides with the open end of the pipe. It is observed that the antinode actually occurs a little above the open end. A correction is applied for this which is known as end a correction. This is denoted by e.

For closed pipe, l is replaced by $l + e = l + 0.3 D$, where D is the diameter of the tube.

For open pipe, l is replaced by $l + 2e = l + 0.6D$, where D is the diameter of the tube.

21

Musical Sound and Doppler Effect

1. Intensity

The intensity of sound at a point in a medium is the average rate at which the energy is transported by the wave, per unit area, across a surface perpendicular to the direction of propagation i.e. Average power transported per unit area. Intensity depends upon the following :

(i) directly proportional to the square of amplitude of the wave,

(ii) directly proportional to the size of vibrating body,

(iii) directly proportional to the density of medium in which sound waves travel,

(iv) inversely proportional to the square of the distance in which sound waves travel,

(v) inversely proportional to the square of the distance from the sounding body.

Remember that loudness of the sound is the degree of sensation produced in the ear.

The relation between loudness L and intensity I is given below

$$L = k \log I$$

where k is a constant.

2. Quality

The quality of sound is that characteristic which enables us to distinguish between different musical notes emitted by different musical instruments. The musical sound for a musical instrument consists of fundamental and many harmonics. Two notes of the same frequency may have harmonics with different intensity distribution. So they are said to differ in quality.

3. Pitch or Frequency

Pitch is the characteristics of sound which distinguishes between a shrill sound and a grave sound. Actually, pitch of a note is the sensation conveyed to our brain by the sound waves falling on our ear that depends directly on the frequency of the incident sound waves. Frequency of a note is a physical quantity and can be measured accurately while the pitch of a note is merely the mental sensation experienced by the observer.

4. Reflection of Sound : Echoes

When a sound wave meets with a large plane obstacle, the wave is reflected. This is because of the fact that sound is a wave motion. Sound waves obey approximately the usual laws of reflection of light. The common experience of sound reflection is echoes heard in large halls and in the neighbourhood of hills. An echo is a repetition of speakers own words caused by reflection at a distance surface e.g., a cliff, a row of building etc. If a shot is fired at a distance d metres from the cliff and echo is heared after t seconds, the velocity of sound in air

$$v = 2b/t.$$

5. Harmonics

The term *harmonics* refers to a series of frequencies which have whole number ratios. If n is the lowest (**fundamental**) frequency called the first harnomic, then the second harmonic has frequency $2n$, third $3n$, fourth $4n$, and so on.

6. Doppler Effect

The apparent change in the frequency or wavelength due to motion of the source relative to the listener or *vice versa* is called Doppler Effect.

(a) **Source in motion and listener at rest** : If the source is approaching the listener with speed 'a'

$$n' = \frac{V}{V - a} \, n \text{ Hertz}$$

where n' is the apparent frequency, and n is the actual frequency.

If the source is receding from the listener with speed a,

$$n' = \frac{V}{V + a} n.$$

As the source passes the listener, the drop in frequency is

$$n \left\{ \frac{V}{V - a} - \frac{V}{V + a} \right\} = \frac{2aV}{V^2 - a^2} \, n.$$

(b) Listener in Motion and Source at Rest : If the listener is moving towards the source with speed b.

$$n' = \frac{V+b}{V} n.$$

If the listener is moving away from the source with speed b,

$$n' = \frac{V-b}{V} n.$$

The fall in frequency as the listener passes the source is

$$\frac{2b}{V} n.$$

(c) Source and Listener both in motion : Let the velocities a and b be considered positive if the motion is from left to right.

$$n' = \frac{V-b}{V-a} n.$$

7. A General Expression for Doppler Effect

A general expression for observed frequency when both listener and observer are in motion or any one of them is in motion is given by

$$n = \left(\frac{v \pm b}{v \pm a}\right) n.$$

Motion of the source affects the denominator of the term to be multiplied by n whereas the motion of the listener is taken care of by the numerator of the same term.

8. Sound Intensity

Intensity of Energy of a beam of sound waves is defined as the energy falling per unit area per unit time, are being kept perpendicular to the direction of propagation of the waves. On calculation it comes out to be,

$$I = 2\pi^2 A^2 n^2 cs$$

where A is the amplitude, n is the frequency, c is the velocity of the waves and s is the density of the medium. It is measured in watts/metre^2.

A practical unit for intensity level measurement is *Decibel*. Intensity level in decibels is equal to ten times the logarithm to base ten of the vatro of its intensity to the standard reference intensity. The standard reference intensity to chosen to have different values when working in different references

Intensity in Decibels

$$= 10 \times \log_{10}\left(\frac{I}{I_0}\right)$$

where I is the intensity to be measured and I_0 is the reference frequency.

9. Diatonic Scale

It is a musical scale consisting of light notes in the following of frequency ratios

C   D   E   F   G   A   B   C
Sa   Re   Ga   Ma   Pa   Da   Ni   Sa

Frequency Ratio

$$1 : \frac{9}{8} : \frac{5}{4} : \frac{4}{3} : \frac{3}{2} : \frac{5}{3} : \frac{15}{8} : 2.$$

22

Reflection and Refraction of Light

1. Basic Terms

(i) **Light.** It is that part of radiant energy from a hot body which produces sensation of vision on human eye. It is a wave motion travelling at a speed of 1,86,000 miles per second or 3,14,000 km/sec.

(ii) **Reflection.** When light falls on a surface, it is divided into three portions. One part returns in that medium in a particular direction. Another part is absorbed and a third part passes through the surface. The light, which is sent back in a particular direction, is called *reflected light* and the phenomenon is known as reflection of light. It obeys the following laws : (i) The incident, ray, reflected ray and the normal to the point of incidence, all lie in the same plane. (ii) The angle of incidence is equal to the angle of reflection.

(iii) **Plain mirrors.** In plain mirrors, image is always behind the mirror, erect, virtual, same size as the object, and at the same distance the mirror, as the object.

Definitions and Formulae in Physics

(iv) **Optical medium.** It is a substance through which light can pass, e.g. air, water, glass etc. It is of three kinds.

(i) Transparent (ii) Translucent (iii) Opaque.

2. Properties of Image Formed by Plane Mirror

1. The image formed by a plane mirror is virtual, erect and laterally reversed.

2. The size of the image is same as the size of the object.

3. The image is as far behind the mirror as the source in front of it.

4. When the plane mirror is rotated through certain angle, the reflected ray turns through double the angle.

5. When two plane mirrors are kept facing each other at angle θ and an object is placed between them, multiple image of the object are formed as a result of multiple successive reflections.

If $(360/\theta)$ is even, then the number of images is given by

$$n = (360/\theta) - 1$$

If $(360/\theta)$ is odd, then the following two situations arise :

(i) if the object lies symmetrically, then

$$n = (360/\theta) - 1$$

(ii) if the object lies unsymmetrically, then

$$n = (360/\theta).$$

When the two plane mirrors are placed parallel to each other, then

$$n = (360/\text{zero}) = \infty \text{ (infinite number of images).}$$

3. Refraction of Light

When a ray of light passes through one medium to another

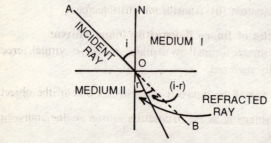

there is a change in the direction of the ray. This is known as refraction. the refracted ray bends towards the normal [Fig.] when the second medium is denser than the first medium and vice-versa. It has been experimentally observed that the velocity of light in medium second also changes. The deviation D suffered by the refracted ray is given by

$$D = |\,i - r\,|.$$

Refraction of light. Laws

(i) The incident ray, the normal and the refracted ray lie in the same plane.

(ii) The ratio of sine of angle of incidence to that of the angle of refraction is constant for the same pair of media. This ratio is called *refractive index* μ,

$$\frac{\sin l}{\sin r} = \mu = \frac{\text{velocity of light in the rarer medium}}{\text{velocity of light in the denser medium}}.$$

If the rarer medium is vacuum (or air), then μ is called the absolute refractive index of the denser medium.

4. Formation of Image by Plane Refraction

The formation of image by plane refraction is shown in

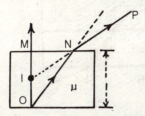

Fig. When an object O, in a denser medium of thickness t and refractive index μ, is seen through a rarer medium, then its image is seen at I. It can be shown that

$$\mu = \frac{\text{real depth}}{\text{apparent depth}} = \frac{MO}{MI}$$

Also, Apparent depth $= (t/\mu)$

and apparent shift $\quad OI = t\left(1 - \dfrac{1}{\mu}\right)$

5. Equivalent Path

Let the velocity of light in a glass slab of thickness t is v. Therefore, its refractive index $\mu = c/v$. Time taken for light to move across the slab is t/v. The distance, light would have travelled in air during this time t/v is called equivalent path and is given be μt.

6. Refraction Through a Number of Media

In Fig. the refraction of a ray of light AB from the medium air to medium water, glass and ultimately to air is shown. Since the first and last medium are the same, it can be shown that the incident ray and emergent ray are parallel. Thus, angle of incidence and angle of emergence are equal.

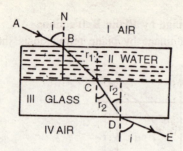

It can be derived that $_a\mu_w \cdot {_w\mu_g} \cdot {_g\mu_a} = 1$

or $\quad _w\mu_g = \dfrac{1}{_a\mu_w \cdot {_g\mu_a}} = \dfrac{_a\mu_g}{_a\mu_w}.$

The apparent shift is given by

$$t_1\left(1 - \frac{1}{\mu_1}\right) + t_2\left(1 - \frac{1}{\mu_2}\right) + \ldots$$

(i) **Critical angle.** Consider a ray travelling from denser medium to rarer medium. The angle of incidence for which the angle of refraction becomes 90° is called the critical angle. In this case

$$\frac{\sin C}{\sin 90°} = \frac{\mu_1}{\mu_2} = \frac{1}{_1\mu_2}.$$

(ii) **Total internal reflection.** When a ray is going from a denser medium to a rarer medium, the ray fails to emerge into the rarer medium if the angle of incidence exceeds a certain critical value. It is then totally reflected in the same medium. This phenomenon is called *Total Internal Reflection*. For this critical angle of incidence θ_c, the ray follows a path along the interface of the two medium.

$$\mu = 1/\sin \theta_c.$$

7. Refraction Through a Prism

The refractive index of the material of the prism can be found by the relation

$$\mu = \sin\left(\frac{A + \delta}{2}\right) / \sin\left(\frac{A}{2}\right).$$

where A is the refracting angle of the prism and δ is the angle of minimum deviation.

8. Real Image

When a beam of light diverging from a point after reflection (or reflection) actually converges to a second point, then the second point is called the real image of the first point. The real image can be formed on the screen.

9. Virtual Image

When a beam of light diverging from a point after reflection (or refraction) appears to diverge from a second point, then the second point is called as the virtual image of the first. The virtual image cannot be formed on the screen.

When a **plane mirror** is rotated through any angle, the **reflected ray** is rotated through twice the angle.

In case of a **plane mirror** the image is as far behind the mirror as the source is in front of it.

When two **plane mirrors** are kept facing each other at an angle θ and an object is placed between them, multiple images of the object are formed as a result of multiple successive reflection. If $\frac{360}{\theta}$ is even, then number of images is given by $n = \left(\frac{360}{\theta} - 1\right)$

23

Image Formation by Curved Mirrors and Lenses

1. Spherical Mirrors

A spherical mirror is a reflecting surface which forms a part of the sphere. When the reflection takes place from the inner surface and outer surface is polished or silvered, the mirror is known as *concave mirror*. When the reflection takes place from the outer surface and the inner surface is polished or silvered, the mirror is known as convex mirror.

Principal focus is defined as the convergence point for rays parallel to the axis of mirror. It is located half way between the mirror and the centre of curvature.

2. Images Formed by Concave Mirror

(i) *Object at infinity.* Image at focus, inverted, real and diminished.

(ii) *Object between infinity and centre of curvature.* Image lies between the focus and the centre of curvature, inverted, real and diminished.

(iii) *Object at centre of curvature.* Image at the centre, inverted, real and of the same size as object.

(iv) *Object between focus and the centre.* Image beyond the centre, inverted real and enlarged.

(v) *Object at the focus.* Image at the Infinity, cannot be seen.

(vi) *Object between mirror and the focus.* Image behind the mirror, erect, virtual and enlarged.

3. Mirror Formula

For all spherical mirrors

$$\frac{1}{u} + \frac{1}{v} = \frac{2}{r} = \frac{1}{f} \text{ as } r = 2f$$

where, u = distance of the object from the pole of mirror
v = distance of the image from the pole of mirror
f = focal length of the mirror
r = radius of curvature of the mirror

Magnification

$$m = \frac{v}{u} = \frac{v-f}{f} = \frac{f}{u-f}$$

or $$m = I/O$$

where I = size of the image and O = size of the object.

Refraction at a Single Spherical Surface

The refraction formula is given by

$$\frac{_1\mu_2}{v} + \frac{1}{u} = \frac{_1\mu_2 - 1}{R}$$

where $_1\mu_2$ is the refractive index of medium second with respect to medium first. If μ_1 and μ_2 be the refractive indices of medium first and second respectively, then above equation can be written as

$$\frac{\mu_1}{u} + \frac{\mu_2}{v} = \frac{\mu_2 - \mu_1}{R}.$$

This formula is applicable for convex as well as concave surfaces. The important point is that u, v and R should be given correct signs.

4. Lens

It is a portion of a transparent medium bound by two spherical surfaces. It may be convex or concave.

Optical centre. It is a point on the principal axis fixed in a position with respect to the lens, such that the direction of every ray incident on the lens which becomes parallel to its direction on emergence must pass through this point.

The *Principal Focus* of a thin lens with spherical faces is the point where rays parallel to the axis are brought to focus. This focus is real for a convex lens and virtual for a concave lens. The focal length f is the distance between the lens and principal focus.

5. Images Formed by Convex Lens

(i) *Object at infinity* : Image at the principal focus on the other side of the lens, inverted real and diminished.

(ii) *Object between 2f and infinity* : Image between f and $2f$ on the other side of the lens, inverted, real and diminished.

(iii) *Object at 2f* : Image at $2f$ on the other side of the lens, inverted, real and of the same size as the object.

(iv) *Object between f and 2f* : Image beyond $2f$ on the other side of the lens, inverted real and enlarged.

(v) *Object at the Principal focus* : Image at infinity, real, inverted and enlarged but the image cannot be seen.

(vi) *Object between the lens and the Principal focus* : Image on the same side of the lens as the object, virtual, erect and enlarged.

6. Image Formed by a Concave Lens

In all cases the image is located on the same side of the lens as the object and it is virtual erect and diminished.

Formula $\dfrac{1}{v} - \dfrac{1}{u} = \dfrac{1}{f}$ for both type of lenses.

Here too the object is always to be placed to the left of the lens. All distances to the left of the lens are negative and those to the right are positive.

$$\text{Linear magnification} = \frac{\text{length of image}}{\text{length of object}} = \frac{v}{u}$$

Lens maker's formula

$$\frac{1}{f} = (\mu - 1)\left(\frac{1}{R_1} - \frac{1}{R_2}\right)$$

where μ = refractive index of the material of the lens, R_1 and R_2 are radii of curvature of the two surfaces of the lens and f is the focal length.

The radius of curvature is positive for convex surface and negative for concave surface.

7. Power of a Lens

It is given by

$$= \frac{1}{\text{focal length (in metres)}} = \text{Dioptre.}$$

The power is positive for convex lenses and negative for concave lenses.

8. Combination of Lenses

(i) When two lenses of focal length f_1 and f_2 are placed in contact, the focal length f of the combination is given by the equation

$$\frac{1}{f} = \frac{1}{f_1} + \frac{1}{f_2}$$

Also, power of the combination = sum of powers of the individual lenses.

(ii) When two lenses of focal lengths f_1 and f_2 are placed coaxically such that the distance between them is d, then the focal length f of the combination is given by the equation

$$\frac{1}{f} = \frac{1}{f_1} + \frac{1}{f_2} - \frac{d}{f_1 f_2}$$

9. Lens Situated in a Liquid

When a lens of refractive index μ_g is placed in a liquid of refractive index μ_1, then the focal length of the lens is given by

$$\frac{1}{f} = \frac{1}{u} + \frac{1}{v} = \left(\frac{\mu_g}{\mu_1} - 1\right)\left(\frac{1}{R_1} - \frac{1}{R_2}\right)$$

or

$$\frac{1}{f} = \frac{\mu_g - \mu_1}{\mu_1}\left(\frac{1}{R_1} - \frac{1}{R_1}\right)$$

10. Silvering at One Surface

When one surface of a thin lens is silvered, then the focal length F of effective lens is expressed as

$$\frac{1}{F} = \Sigma \frac{1}{f_l} \qquad \ldots(1)$$

where f_l is the focal length of the lens or mirror to be repeated as many times as the refraction or reflection respectively is repeated.

(i) *Focal length of planoconvex lens when silvered at its plane surface.* When an object is placed in front of such a lens, the rays first of all refracted from the convex surface, then refracted from the polished plane surface and again refracts out from convex surface. If f_l and f_m be the focal lengths of lens (convex surface) and mirror (plane polish surface) respectively then effective focal length of the lens F is given by

$$\frac{1}{F} = \frac{1}{f_l} + \frac{1}{f_m} + \frac{1}{f_l} = \frac{2}{f_l} + \frac{1}{f_m} \qquad \ldots(2)$$

Here $\quad \dfrac{1}{f_l} = (\mu - 1)\left(\dfrac{1}{R}\right)$ and $f_m = \dfrac{R}{2} = \infty$

$\therefore \qquad \dfrac{1}{F} = \dfrac{2}{f_l}$ or $F = \dfrac{f_l}{2} \qquad \ldots(3)$

(ii) *Focal length of planoconvex lens when silvered at convex surface.* In this case

$$\frac{1}{F} = \frac{1}{f_l} + \frac{1}{f_m} + \frac{1}{f_l}$$

$$= \frac{2}{f_l} + \frac{1}{f_m} = \frac{2}{f_l} + \frac{1}{(R/2)}$$

$$= 2\left[\frac{(\mu - 1)}{R}\right] + \frac{2}{R} = \frac{2\mu}{R} \qquad \because \frac{1}{f_l} = \frac{(\mu - 1)}{R}$$

or $\quad F = (R/2\mu) \quad$ ...(4)

(iii) *Focal length of convex lens whose convex surface of radius R_2 is silvered.* In this case

$$\frac{1}{F} = \frac{1}{f_l} + \frac{1}{f_m} + \frac{1}{f_l} = \frac{2}{f_l} + \frac{2}{R_2} \quad ...(5)$$

24

Defects of Eye and Optical Instruments

1. Defects of Vision

(i) *Short-sightedness (Myopia).*

The defect is corrected by using concave lenses in the spectacles. Focal length of the lens is suitably chosen.

(ii) *Long-sightedness (Hypermetropia).*

To correct long-sightedness one should use convex lenses in the spectacles and the focal length should be suitably chosen.

2. Optical Instruments

Simple microscope or magnifying glass

$$Magnifying\ power = 1 + \frac{D}{f}$$

[where D is the least distance of distinct vision]

Compound microscope

$$\text{Magnifying power} = \frac{V}{U}\left(1 + \frac{D}{f}\right)$$

[where U is the distance of the object from the objective and V is the corresponding distance of the real image]

Astronomical telescope

For normal adjustment when the final image is formed at infinity.

$$\text{The magnifying power} = \frac{F}{f}$$

[where F and f are focal lengths of the objective and the eyepiece respectively.]

When the final image is formed at D, the least distance of distinct vision

$$\text{The magnifying power} = \frac{F}{f}\left(1 + \frac{D}{f}\right)$$

3. Terrestrial Telescope

It consists of a convex lens objective of long focal length (f_0) and aperture and a convex lens eyepiece of short focal length (f_e) and aperture.

$$\text{Length of the tube of telescope} = (f_0 + f_e) \qquad ...(9)$$

$$\text{Magnifying power} = (f_0/f_e) \qquad ...(10)$$

4. Galilean Telescope

It consists of a convex lens objective of long focal length (f_0) and a concave eyepiece of short focal length (f_e).

$$\text{Length of the tube} = (f_0 - f_e) \qquad ...(11)$$

$$\text{Magnifying power} = (f_o/f_e) \qquad ...(12)$$

5. Resolving Power of Telescope

The resolving power of a telescope is defined as the power to produce distinctly separate images of two close objects. The expression for resolving limit a is given by

$$a = 1.22\,(\lambda/d)$$

where λ is the wavelength of light used and d is aperture (diameter of objective lens).

6. Epidiascope

Epidiascope is the instrument used for projecting the image of opaque objects (such as pictures) on the screen.

7. Projection Lantern

This is used to project a magnitude image of small and transparent objects on the screen.

25

Dispersion of Light and Chromatic Aberration

1. Prism

Prism is a homogeneous, transparent medium (such as glass) that is enclosed by two plane surfaces inclined at an angle. These surfaces are known as the refracting surfaces and the angle between them is known as the angle of the prism. A prism is shown in Fig. (1).

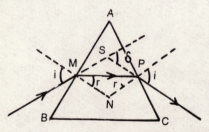

In this $\angle BAC$ is the angle of prism. The angle between the incident ray and the emergent ray is known as the angle of deviation δ.

$$\delta = (\mu - 1) A \qquad \qquad ...(1)$$

where μ is the refractive index of the material of the prism. When the angle of incidence in a prism is increased, the value of angle of deviation δ first decreases and then increases. The minimum value of deviation is known as the *angle of minimum deviation*.

The refractive index m of the material of the prism is given by the following relation.

$$\mu = \frac{\sin(A + \delta_m)/2}{\sin A/2} \qquad ...(2)$$

where δ_m = angle of minimum deviation
and A = angle of the prism.

When ray of light passes through a prism (angle A) in a position of minimum deviation, then the angle of incidence (i) and angle of refraction (r) are given by the relations given below

$$i = (A + \delta_m)/2 \qquad ...(3)$$

$$r = A/2 \qquad ...(4)$$

(i) **Dispersion.** It means the splitting up of a beam of light containing different frequencies by passage through a substance (such as a glass prism) whose refractive index varies with the frequency.

(ii) **Dispersive power** $\omega = \dfrac{\mu_v - \mu_r}{\mu_v - 1}$

where μ_v = refractive index for violent ray
   μ_r = refractive index for red ray
   μ_v = refractive index for the mean ray (yellow).

(iii) **Angular dispersion** in a prism

$$\delta_v - \delta_r + (\mu_v - \mu_r) A$$

(iv) **Mean deviation** $\delta = (\mu - 1) A$

where δ_v = deviation for the violet ray
    δ_r = deviation for the red ray.

Combination of two prisms to produce deviation without dispersion

Condition : $w_1 \delta_1 = - w_2 \delta_2$

or $w_1(\mu_1 - 1) A_1 = - w_2(\mu_2 - 1) A_2$.

Combination of two prisms to produce dispersion without deviation.

Condition : $(\mu_1 - 1) A_1 = - (M_2 - 1) A_2$.

Condition of achromatism of two lenses in contact

$$\frac{w_1}{f_1} = - \frac{w_2}{f_2}.$$

Deviation without Dispersion

By deviation without dispersion we mean an achromatic combination of two prisms where in net or resultant dispersion is zero and deviation is produced. For the two prisms,

$$(\mu_V - \mu_R) A + (\mu'_B - \mu'_R) A' = 0$$

or $A' = \dfrac{(\mu_V - \mu_R)}{(\mu' - \mu'_R)} A$ ...(7)

and $\omega \delta + \omega' \delta' = 0$

where ω and ω' are the dispersive powers of two prisms and δ and δ' are their mean deviation.

Dispersion without deviation

An achromatic combination of two prisms where in the deviation produced for the mean ray by the first prism is equal and opposite to that produced by the second prism is known a direct vision prism. This combination produces dispersion without deviation.

For deviation to be zero, $\delta + \delta' = 0$

or $\quad\quad\quad (\mu - 1) A + (\mu' - 1) A' = 0$

or $\quad\quad\quad A' = -(\mu - 1) A/(\mu' - 1)$ ...(8)

2. Spectrum

It refers to the coloured pattern obtained on the screen after dispersion of light. If there is no overlapping of the colours, it is known as a *pure spectrum*. A spectrometer is a device used for analysing a beam of light into its components colours. It is made up of three parts (i) Collimator, (ii) Prism table and (iii) Telescope. The spectra is classified as under :

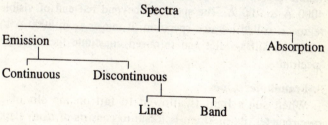

Continuous spectrum. It is made up of a wide range of unseparated wavelengths. Dense hot gases and white hot solids produce continuous spectrum.

Line spectrum. This spectrum is made up of sharp lines of definite wavelengths. Substances in atomic states (sodium vapour lamp, mercury vapour lamps, gases in discharge tubes) give line spectrum.

Bond spectrum. This spectrum is made up of bright bands each having a sharp edge. Incandescent vapours in the molecular state (calcium or barium salts in the bunsen burner flame, nitrogen in molecular state in vacuum tube) give band spectra.

When light from a source is seen through a spectroscope, *emission spectrum* is obtained. On the other hand, when light emitted from a source is made to pass through an absorbing material and then seen through a spectroscope, *absorption spectrum* is obtained.

Visible and Invisible Spectrum

The spectrum that is spread from 4000 Å to 7800 Å and consists of red and violet colours at its each end respectively is called *visible spectrum.* The spectrum beyond violet end of visible region is called ultraviolet spectrum. It ranges from 4000 Å to 100 Å. The spectrum beyond red end of visible region is called infrared spectrum. It ranges from 8000 Å to one mm. Ultra-violet and infra-red constitute the invisible spectrum.

3. Fraunhofer Lines

When sun's light is allowed to fall on the slit of a spectrometer, its spectrum is found to consists of many dark lines. These lines are known as Fraunhofer lines. The number of these lines is about 700. These lines are due to the absorption of certain wavelengths in sunlight by the gases in the sun's outer sphere. It is an example of line absorption spectrum.

4. Fluorescence

The phenomenon of absorption of light of one wavelength by a substance and then re-emission of light of greater wavelength is called fluorescence. Such substances are fluorspar (calcium fluoride), paraffin oil, Quinine sulphate, uranium oxide, etc.

5. Phosphorescence

The phenomenon of re-emission of visible light even after the incident light is cut off is called phosphorescence. Examples are cadmium sulphide, barium sulphide, strontium sulphide, zinc sulphide etc.

6. Rainbow

It is an important spectrum produced by the dispersion of sunlight by the water drops settling down after rain.

7. Chromatic Aberration

When a parallel beam of white light is allowed to pass through a lens, different colours are focussed at different places. This is due to the fact that a lens has different focal lengths for different colours. This defect of a lens is known as chromatic aberration. The chromatic aberration may also be defined in the following manner. The image of a white object formed by a lens is generally coloured and blurred. This defect of the image produced by a lens is known as chromatic aberration.

Chromatic aberration = Dispersive power × mean focal length = $\omega \times f$.

8. Achromatic Doublet

The removal of chromatic aberration is called achromatism. Achromatic doublet is a combination of two lenses adjusted in such a way that the image of a white object is white ie, free from colours. The condition of achromatism is

$$\frac{\omega_1}{f_1} + \frac{\omega_2}{f_2} = 0$$

where ω_1 and ω_2 are the dispersive powers of two lenses respectively.

9. Spherical Aberration

Even by using a monochromatic light, all the rays from any axial point after refraction through a lens are not focussed at a single point due to large aperture of the lens. This defect is known as spherical aberration. This defect can be minimised by using two plano-convex lens separated at a distance d such that $d = f_1 - f_2$.

10. Colours of the Objects

The colours of the object depends upon

(i) the colour of incident light,

(ii) colour of light absorbed by bodies,

(iii) effect of reflected or transmitted light on the retina of eye.

The colours of the opaque objects are due to selective absorption of various colours. The colours of transparent bodies depends upon transmitted light.

11. Sun

These consists of mainly two parts, **Photosphere.** Its temperature is as high as 2×10^7 kelvin, gives continuous spectra. **Chromosphere** surrounds photosphere.

12. Rainbow

It is an important spectrum formed by the dispersion of sunlight by the water drops settling down after rain.

A primary rainbow is violet on the inside and red in outside.

26
Optical Instruments and Miscellaneous Concepts on Light

1. Simple Microscope

The simple microscope is a single double convex lens. This is used to read small divisions of various scales and is also called as reading lens or magnifier. The **magnifying power** M of a microscope is defined as the angle subtended by the image at eye to the angle subtended by the object at eye.

For normal vision, when the final image is formed at infinity, the magnifying power is given by

$$M = D/f$$

where d = distance of distinct vision = 25 cm. It is observed that for every eye there is a least distance at which an object appears to be most distinct. This is called as the least distance of distinct vision.

For distinct vision,

$$M = [1 + (D/f)].$$

2. Compound Microscope

It consists of an objective lens of very short focal length (f_0) and aperture and an eyepiece of comparatively slightly greater focal length (f_e) and aperture.

For image at least distance of distinct vision, the magnifying power is given by

$$M = \frac{v_0}{u_0}\left(1 + \frac{D}{f_e}\right)$$

$$= M_0 \times M_e$$

($\therefore v_0/u_0 = M_0$, magnification produced by object, u_0, v_0 are the distances of object and its image from objective).

When the final image is formed at infinity, the magnifying power is given by

$$M = \frac{v_0}{u_0}\left(\frac{D}{f_e}\right)$$

3. Astronomical Telescope

It consists of an objective lens (convex) or large focal length f_0 of large aperture and a convex lens eyepiece of very short focal length (f_e) and small aperture.

The optical length of tube of telescope $= (f_0 + f_e)$

The magnifying power is given by

$$M = \frac{f_0}{f_e}\left(1 + \frac{f_e}{v_e}\right)$$

If the final image is formed at a distance D, then the magnifying power is given $M = \dfrac{f_0}{f_e}\left(1 + \dfrac{f_e}{D}\right) v_e = D$.

If the final image is formed at infinity, then the magnifying power is given by

$$M = \dfrac{f_0}{f_e} \quad (v_e = \infty).$$

4. Terrestrial Telescope

It consists of a convex lens objective of long focal length (f_0) and aperture and a convex lens eyepiece of short focal length (f_e) and aperture.

Length of the tube of telescope = $(f_0 + f_e)$

Magnifying power = (f_0/f_e).

5. Galilean Telescope

It consists of a convex lens objective of long focal length (f_0) and a concave eyepiece of short focal length (f_e).

Length of the tube = $(f_0 - f_e)$

Magnifying power = (f_0/f_e).

6. Resolving Power of Telescope

The resolving power of a telescope is defined as the power to produce distinctly separate images of two close objects. The expression for resolving limit a is given by

$$a = 1.22\,(\lambda/d)$$

where λ is the wavelength of light used and d is aperture (diameter of objective lens).

7. Epidiascope

It is the instrument used for projecting the image of opaque objects (such as pictures) on the screen.

8. Projection Lantern

This is used to project a magnified image of small and transparent objects on the screen.

Sextant is used to measure the angle subtended at the eye by two distant objects.

$$h = \frac{d}{\cot \theta_1 - \cot \theta_2}$$

where θ_1 and θ_2 are the angles of elevation at the two places at a distance 'd' apart each other.

9. Defects of Vision

Following are defects of vision :

(i) *Myopia (short sightedness)*. In this defect, i.e., eye cannot see the distant objects clearly (can see the nearer objects clearly). This defect can be removed by using a concave lens of suitable focal length.

(ii) *Hypermetropia (long sightedness)*. In this defect, the eye cannot see the nearer objects clearly (can see the distinct objects clearly). This defect can be removed by using a convex lens of suitable focal length.

(iii) *Presbyopic Eye* can see distant objects clearly but not near objects. This defect is due to hardening of the crystalline lens which loses elasticity gradually all through life. Power of accomodation decreases. A Presbyopic eye can see between 50-100 cm.

Definitions and Formulae in Physics

(iv) *Astigmatism.* An eye is said to have this defect when it cannot see simultaneously with the same distinctness horizontal and vertical lines at the same distance. It can be removed by using cylindrical or spherocylinderical lenses.

10. Velocity of Light

In vacuum or space, the velocity of light is 3×10^8 m/sec. It is the maximum possible velocity which a physical particle can approach. The velocity of light by different methods is given by :

(i) **Fizeau's Method.** According to this method, the velocity of light

$$c = 4\, nmd.$$

n = number of revolutions/sec made by the wheel when the first eclipse occurs.

m = number of teeth

d = distance between the farther wheel and the distant mirror

The second and third eclipses occur when the speed of rotation is 3 and 5 times respectively. In general, the velocity of light is given by

$$c = \frac{4\, n\, m\, d}{(2p - 1)}$$

where p = number of eclipses that have occurred.

(ii) **Focault's rotating mirror method.** According to this method, the velocity of light c is given by

$$c = \frac{8\pi n a r^2}{(b + r)x}$$

where r = distance between plane mirror and concave mirror

a = distance of the object from achromatic lens

$(b + r)$ = distance of the image from achromatic lens

(iii) **Michelson's method.** According to this method, the velocity of light

$$c = 8\,n\,d$$

where n = number of rotations made by the octagonal mirror per second

d = distance travelled by light.

11. Newton's Corpuscular Theory

According to this theory :

(i) Light consists of extremely small, very light weighed material particles called 'corpuscles'.

(ii) The corpuscles travel with the velocity of light.

(iii) When the corpuscles strike our retina, they produce the sensation of vision.

(iv) Corpuscles of different colours are of different sizes.

This theory explains the reflection, refraction, rectilinear propagation of light, etc., but fails to explain the phenomena of interference, diffraction and polarization of light.

12. Wave Front

The locus of all particles vibrating in the same phase is called wave-front.

13. Huygen's Wave Theory

In 1679, Huygen's proposed the wave theory of light. According to wave theory of light, each point in a source of light sends out waves in all directions in hypothetical medium called ether. Ether was assumed to be continuous medium which pervades all space. The existence of ether had to be assumed because for the propagation of wave motion, a medium is necessary.

14. Huygen's Principle

This principle provides a geometrical method of finding the shape and position of the wavefront at a certain instant from its shape and position of some earlier. The principle is stated in the following two parts :

(i) Each point on the wavefront acts as a centre of new disturbance and emits its own set of spherical waves called secondary wavelets. The secondary wavelets travel in all directions with the velocity of light so long as they move in the same medium.

(ii) The envelope or the locus of these wavelets in the forward direction gives the position of new wavefront at any subsequent time.

The principle explained successfully the reflection, refraction interference, diffraction etc., but failed to explain the rectilinear propagation of light.

15. Maxwell's Electromagnetic theory

According to Maxwell's electromagnetic theory, the light wave is composed of an oscillating electric field combined with a magnetic field perpendicular to it and travelling with the velocity of light perpendicular to both.

16. Planck's Quantum Theory

According to Planck's quantum theory the light moves in the form of small bundles or packets called photons. The energy of each photon is $h\nu$ where ν is the frequency of light and h is planck's constant.

17. Modern Theory

According to modern theory, the light consists of both particles and wave properties, i.e., a dual nature. The wave properties are manifested in the phenomena of interference, diffraction and polarisation. On the other hand, light is a stream of particles called photons which travel with the velocity of light.

18. Interference of Light

When the two waves of same frequency with zero initial phase difference or constant phase difference superimpose over each other, then the resultant amplitude (or itensity) in the region of superimposition is different than the amplitude (or intensity) of individual waves. *The modification in intensity in the region of superimposition is called interference.* When the resultant amplitude is sum of the amplitudes due to two waves, the interference is known as *constructive intereference* and when the resultant amplitude is equal to the difference of two amplitude, the intereference is known as *destructive intereference.* There is no violation of the law of conservation of energy in intereference. Here the energy from the points of minimum energy is shifted to the points of maximum energy.

19. Conditions for Interference of Light

To obtain the stationary intereference pattern, the following conditions should be satisfied :

(i) The two sources should be coherent i.e., they should vibrate in the same phase or there should be a constant phases difference between them.

(ii) The two sources must emit continuous waves of the same wavelength and time period.

(iii) The separation between two sources ($2d$) should be small.

(iv) The distance D between two sources and screen should be large.

(v) The amplitudes of the intereferring waves should be equal.

(vi) Sources should be narrow.

(vii) Sources should be monochromatic.

20. Methods of Producing Coherent Sources

There are two general methods of producing coherent sources :

(i) *Division of wave-front.* In this method, the wavefront is divided into two or more parts with the help of mirrors, lenses and prisms. The common methods are Young's double slit arrangement, Fresnel's biprism method, Lioyd's mirror method, etc.

(ii) *Division of amplitude.* In this method, the amplitude of the incoming beam is divided into two or more parts by partial reflection or refraction. These divided parts travel different paths and finally brought together to produce interference. Newton's rings, Michelson's interferometer, etc., are the examples of division of amplitude. The common example is the brilliant colours observed when a thin film of transparent material like soap bubbles or thin film of kerosine oil spread on the surface of water is exposed to an extended source of light.

Fringe-width. The distance between any two consecutive fringes (bright or dark) is known as fringe-width. This is denoted by β and is given $\beta = (\lambda D/2d)$, where λ is the wavelength of light used, D is the distance of the screen from the sources and $2d$ is the separation between two coherent sources.

Displacement of fringes. When a thin transparent plate of thickness t and refractive index μ is introduced in the path of one of the interfering wave then the entire fringe pattern is shifted through a constant distance, of course, fringe width remains the same.

If S be the shift in the nth maximum fringe, then

$$S = \frac{D}{2d}(\mu - 1)t.$$

21. Diffraction of Light

When light falls on obstacles or small apertures whose size is comparable with the wavelength of light used, then there is a departure from straight line propagation, the light bends round the corners of the obstacles or apertures. *The bending of light round the corners of an obstacle or aperture is called diffraction.* The diffraction is a common phenomenon to all wave motion. The diffraction can easily be observed in case of sound waves because the wavelength of sound waves is too much.

Considering the diffraction at a single slit, the minimum intensity positions are given by

$$e \sin \theta \pm \neq m \lambda$$

where $m = 1, 2, 3, \ldots$ etc.
       e = width of the slit
       θ = angle of diffraction

22. Unpolarised Light

The ordinary light also called as unpolarised light, consists of a very large number of vibrations in all planes with equal probability at right angles to the direction of propagation. So unpolarised light is symmetrical about its direction of propagation.

23. Polarised Light

The light which has acquired the property of one-sideness is called *polarised light* or the lack of symmetry of vibration around the direction of wave propagation is called *polarisation*. Polarisation of light waves exhibits that they are transverse waves. When the vibrations are confined only to a single direction in a plane perpendicular to the direction of propagation, it is called a *plane polarised light*. A plane passing through the direction of propagation and perpendicular to the plane of vibration is known as *plane of polarisation*.

24. Production of Plane Polarised Light

The plane polarised light can be produced by the following methods :

(a) by reflection : Brewster's law $\mu = \tan p$, where p is the angles of polarisation (Brewster's angle) and μ is refractive index of the reflecting medium.

(b) by refraction (piles of plates)

(c) by dichroism

(d) by double refraction (Nicol's prism)

(e) by scattering.

25. Nicol's Prism

Nicol's prism is device for producing and analysing a plane polarised light. It is based on the phenomenon of double refraction. Bartholinus discovered that when a beam of ordinary unpolarised light is passed through a calcite crystal, the refracted light is split up into two refracted rays. The one which always obeys the ordinary laws of refraction and having vibrations perpendicular to the principal section is known as ordinary ray. The other, in general, does not obey the laws of refraction and having vibrations in principal section is called as extra-ordinary ray. Both the rays are plane polarised. The phenomenon is known as *double refraction*. The crystals showing this phenomenon are known as doubly refracting crystals.

26. Polaroids

Polaroids are artificially made polarising materials in the form of sheets or plates capable of producing strong beam of plane polarised light.

27. Spectrometer

It is a device for analysing a beam of light into its component colours. It consists of three parts.

(i) Collimator (ii) Prism table (iii) Telescope.

Direct Vision Spectroscope. It is an instrument in which spectrum is conveniently obtained in the direction of incident light. It disperses the light without producing any mean deviation.

28. Photometry

It is the branch of physics in which the power of emitting light energy of a light source and illumination produced on the surface is measured.

29. Radiant Flux

It is the amount of radiant energy which flows per second through a normal section of beam. Its unit is watt.

Definitions and Formulae in Physics

$$\text{Radiant flux} = \frac{\text{Joule}}{\text{Sec}} = \text{watt.}$$

30. Luminous Flux

It refers to visible radiant energy emitted by a source per second. Its unit is **Lumen.**

31. Solid Angle, $\Delta\omega$.

$$\Delta\omega = \frac{\Delta A \cos\theta}{r^2}$$

where ΔA = Area of the element,
 r = distance of middle point from Q,
 θ = angle between positive direction of normal at spherical surface.

32. Luminous Intensity

The luminous intensity I of a point light source in any direction is defined as the luminous flux per solid angle emitted by the source in that direction

$$I = \Delta F/\Delta\omega$$

where ΔF is the luminous flux emitted by light source in solid angle $d\omega$.

The unit of the luminous intensity is lumen/steradian. This is also called as 'candela'

$$\text{Total flux } F = 4\pi I \text{ lumen.}$$

33. Brightness of a Surface

If ΔF be the luminous flux falling on an area ΔA, then brightness or intensity of illumination is given by

$$E = \frac{\Delta F}{\Delta A}$$

Its unit is *Lux*, 1 Lux = $1 \dfrac{\text{lumen}}{\text{metre}^2}$.

34. Lambert Cosine Law

$$E = \frac{I \cos \theta}{r^2}$$

or
$$E \propto \cos \theta$$

Inverse square law states that the intensity of illumination at a point due to a given source varies inversely as the square of the distance of the point from the source.

27

Magnetism

1. Lines of Force of a Magnetic Field

These are imaginary lines along which a north pole would move if free to do so in a magnetic field and are such that tangents to them at any point give the direction of magnetic field at that point. Some of its important properties are

(i) The lines of force never cut each other.

(ii) They start from N-pole, move along smooth curve upto the S-pole.

(iii) The tangent at any point of the line of force represents the direction of the resultant intensity at the point.

(iv) The tangent at any point of the line of force represents the direction of the resultant intensity at that point.

(v) The number of magnetic lines of forces through unit area (normally) is equal to the intensity of magnetic field at that point.

(vi) In a uniform magnetic field, lines of force are parallel straight lines while in a non-uniform field they are curve crowding at the poles.

(vii) A natural point in a magnetic field is the point where no line of force exists. It is a point in a magnetic field where resultant magnetic intensity is zero.

2. Molecular Theory of Magnetism

It states that (i) molecules of magnetic substances are complete magnets in themselves; (ii) when a substance is magnetised molecular magnets are aligned in the direction of magnetising field; (iii) magnetism is due to the revolution and rotation of the electrons within the atom.

Magnetic field. Any region or space around a magnet in which its magnetic effects can be detected is referred to as 'magnetic field'.

Intensity. The force exerted on a unit north pole to a point in a magnetic field is called the intensity of the magnetic field at that point. The unit in C.G.S. system is *oersted* and in M.K.S. it is ampere turns/pole

$$(= 4\pi \times 10^{-3} \text{ oersted}).$$

Unit magnetic pole. A pole has a unit strength. If it is repelled with a force of a 1 dyne when placed at a distance of 1 cm in air from an equal and similar pole.

Pole strength of a magnetic pole is numerically equal of the force it exert, on a unit pole placed in air at a distance of 1 cm from it.

Uniform magnetic field. It has the same insensity at all points situated in it and can be represented by parallel lines of force.

Magnetic moment. A magnet is the moment of the couple acting on a magnet when it is kept perpendicular to a uniform field of unity intensity.

Definitions and Formulae in Physics

Inverse square law. The force of attraction or repulsion between two magnetic poles is directly proportional to their pole strengths and inversely proportional to the square of distance between them.

Thus
$$F = \frac{1}{\mu} \frac{m_1 m_2}{r^2}$$

where F is the magnetic force between two magnets of strength m_1 and m_2 separated by a distance r in medium of permeability μ. For air $\mu = 1$.

Tangent law. If a magnet be acted upon by the uniform and mutually perpendicular fields, it will remain in equilibrium at an angle θ with one of the field such that the tangent of the angle θ is equal to the ratio of two fields.

$$F = H \tan \theta$$

The intensity of magnetic field at a point in the end of position. The field due to a magnet of magnetic moment M of length $2l$ at a point in the end on position at a distance d is

$$F = \frac{2Md}{(d^2 - l^2)^2} \text{ and if } d^2 >> l^2 F = \frac{2M}{d^2}.$$

The intensity of magnetic field at a point in the broad side on position. The field due to magnet is given by

$$F = \frac{M}{(d^2 + l^2)^{3/2}} \text{ and if } d^2 << l^2 . F = \frac{M}{d^3}.$$

Magnetic potential. It is the amount of work done in bringing a unit north pole against the magnetic intensity from an infinite distance upto the point at which the potential is to be determined.

Magnetic shell. A magnetic shell is a thin sheet of magnetic material magnetised at every point in a direction perpendicular to the surface of the shell. A shell can be of any shape. One face of the shell is entirely a north pole and the whole of the other face is a south pole.

Deflection magnatometer. It is an instrument in which a small pivoted magnetic needle obeys the tangent law, $F = H \tan \theta$. Deflection magnetometer can be set in either of the two positions

(i) Tangent A position

(ii) Tangent B position

It utilises either deflection method or null method. It is used for (i) comparison of magnet moments; (ii) comparison of field due to magnets; (iii) comparison of magnetic field at two places; and (iv) verification of Gauss' inverse square law.

3. Vibration Magnetometer

It consists of a box in which a small bar is freely suspended by an unspun silk thread. The magnet can be execute torsional vibrations about thread as an axis in a horizontal plane and its time period $t = \pi \sqrt{(I/MH)}$ where I = Moment of Inertia of the magnet about the thread. It is used for (i) comparison of magnetic moments; (ii) comparison of field due to the moments; (iii) determination of M or H by using both deflection and vibration magnetometers.

4. Magnetic Elements

The elements that characterise the earth's magnetism at a place are : (i) declination, (ii) Dip, (iii) horizontal component of earth's magnetic field.

Definitions and Formulae in Physics

(i) **Declination.** It is the angle between the magnetic meridian and geographical meridian.

(ii) **Dip.** It is the angle between the direction of the total intensity of earth's magnetic field and the horizontal direction in the magnetic meridian. It is denoted by δ.

(iii) **Horizontal component of earth's field.** A place is the component of the intensity of the earth's magnetic field along a horizontal place. It is denoted by H.

Dip circle. It is an instrument to determine the magnetic dip at a place. The value of dip increases suddenly when we approach a locality of iron ore. Thus a dip circle may be used for detecting beds of iron ores.

Isogonals or isogonic lines. These are imaginary lines joining points of equal declination. The imaginary lines joining to point of equal dip are called *isoclinic lines*. The lines having same value of H are called *isodynamic lines*. The value of magnetic elements changes and changes are of different types-secular, annual, lunar and daily. All changes occur gradually.

28
Properties of Magnetic Materials

1. Basic Terms

Magnetic materials. These are materials in which a state of magnetisation can be induced. When such materials are magnetised they create a magnetic field in the surrounding space.

Intensity of magnetisation. It is the ratio of magnetic moment developed in the specimen to its volume. It is denoted by I. ($I = M/V$).

Diamagnetic substances. These are substances which are repelled by magnets and when placed in a magnetic field move from stronger to weaker part of the field.

Ferromagnetic substances. These are substances which when placed in magnetic field acquire a strong magnetisation in the direction of field.

Curie-temperature. The temperature above which a ferromagnetic substance becomes paramagnetic is called as "curie-temperature" of the substance. Curie-temperature of iron is 770°C and that of nickel of 358°C.

2. Retentivity, Coercivity and Hysteresis Loop

If an unmagnetised bar of a ferro-magnetic substance is placed in an increasing magnetic field H, then it is seen that the intensity of magnetisation I_m (or B, the magnetic induction) first increases slowly and then rapidly as shown in Fig. (1) till it becomes constant. This value of intensity of magnetisation is known as *saturation value* (denoted by point A). The substance is then said to be saturated. If we increase the magnetising force beyond the point A, I_m remains constant.

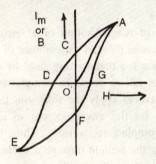

Fig. 1.

Retentivity. After reaching the point A, if H is gradually decreases, then it is found that the curve does not coincide with OA but takes the form AC. This show that when the value of H is zero, the intensity of magnetisation (or magnetic induction B) is not zero but has a value OC. This value of $I_m = (OC)$ is known as *residual magnetism or remanence or retentivity*.

Coercivity. To destroy the residual magnetism, we reverse the direction of magnetising field H. It is observed that for

$H = OD$, the intensity of magnetisation is zero i.e., $B = 0$. The value of H needed to destroy the residual magnetism is known as the coercivity.

Hysteresis loop. On further increasing H in the reverse direction, the curve DE is obtained. The saturation point E is symmetrical to point A. Now H is increased in steps and then we reach the point A by following the path $EFGA$. Thus in all cases, I_m appears to lag behind the magnetising field. This lagging of intensity of magnetisation behind the magnetising field is known as hysteresis. The path $ACDEFGA$ is known as hysteresis loop.

3. Hysteresis Loss

In the process of magnetisation of a ferromagnetic substance through a cycle, there is expenditure of energy. When we gradually reduce the magnetising field to zero, the magnetisation does not fall to zero. To completely demagnetise the sample, we have to apply the magnetic field in the opposite direction. So the energy spent in magnetising a specimen is not complete recoverable. Thus, there is a loss of energy is taking the sample through a cycle of magnetisation. The loss of energy appears in the form of heat. This loss of energy is known as hysteresis loss.

4. Soft and Hard Materials

The shape of the hysteresis loop is a characteristic of a magnetic material. Fig (2) shows the B-H curve for iron and steel. From the curve, the following conclusions, can be drawn.

(i) Substance having large value of *coercive forces* are magnetically hard, whereas those with small coercive forces are soft.

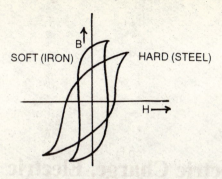

Fig. 2.

(ii) For soft materials, *retentivity* is greater in comparison to that for hard materials.

(iii) The *susceptibility* is greater for soft materials in comparison to for hard materials.

(iv) The *permeability* is greater for soft materials in comparison to for hard materials.

(v) The *hysteresis loss* is less for soft materials in comparison to for hard material.

The hysteresis curve and its properties discussed above provide very useful information for choosing a particular material for practical use in industry.

29

Electric Charge, Electric Field and Potential, Coulomb's Law

1. Electric Charge

Thales a Greek philosopher (640-556 BC) observed that a piece of amber, on being rubbed with fur or cat skin, developed the property of attracting small bits of paper, straw or feathers towards it.

Now it is known that any two substances on being rubbed acquire this attractive property. The bodies are then considered to become *electrified* or to *have acquired electricity* or *charged with electricity*, e.g., when you comb dry hairs, they sometimes produce cracking sound.

2. Two Kinds of Charges

Glass rod rubbed with silk and ebonite rod rubbed with fur shows opposite behaviour, i.e., when a glass rod rubbed with silk in suspended and another glass rod rubbed with silk is brought near it a repulsion is observed however, when ebonite rod rubbed with fur is brought near it an attraction is observed.

Definitions and Formulae in Physics

From this we conclude that these are *two types* of charges. Moreover similar charges repel whereas opposite charges attract each other.

The charge that develops on glass rod on being rubbed with silk is known as *positive* and the one that develops on ebonite rod on being rubbed with fur is known as *negative* charge.

Conservation of charge. Electric charge can neither be produced nor destroyed. When an object loses charge of either kind an equal amount of charge appears elsewhere, e.g. When a glass rod rubbed with silk acquires a positive change the silk used for rubbing also acquires an equal negative charge.

3. Electrostatic Field

The space surrounding a charged body within which the electrostatic forces can be detected is known as *electrostatic field*. Electrostatic field plays the role of an intermediary in the transmission of forces between the charged bodies.

Electric field intensity (field intensity or field strength). The electric field intensity $\vec{E}$ at any point in an electrostatic field is defined as the force exerted on a unit positive charge placed at that point. If a body carrying charge $+q$ when placed at a point in an electric field experiences a force $\vec{F}$, the electric field intensity is given by

$$\boxed{\vec{E} = \frac{\vec{F}}{q}}$$

where $\vec{E}$ is a vector whose direction is the same as direction along which the unit positive charge placed at that point will tend to move if free to do so.

4. Electric Field Due to a Point Charge

Consider a point P, in vacuum, at a distance r from the point charge $+Q$ at point O.

To find the electric field $\vec{E}$ at point P we have to imagine that a test charge Σ is placed at P.

The according to Coulomb's law, we have a force $\vec{F}$ as equal to

$$\vec{F} = \frac{1}{4\pi\varepsilon_0} \cdot \frac{Qq}{r^2} \hat{r}$$

Thus,

$$\vec{E} = \frac{\vec{F}}{q}$$

$$= \frac{1}{4\pi\varepsilon_0} \cdot \frac{q}{r^2} \hat{r} \qquad ...(i)$$

If the absolute permitivity of medium in which charge Q lies is ε_a, then

$$\vec{E} = \frac{1}{4\pi\varepsilon_a} \cdot \frac{Q}{r^2} \hat{r} \qquad ...(ii)$$

The equation (i) and (ii) gives the values of electric intensity in S.I. units.

In c.g.s. units these become

$$\vec{E} = \frac{Q}{r^2} \hat{r} \qquad ...(iii)$$

and

$$\vec{E} = \frac{1}{k} \cdot \frac{Q}{r^2} \hat{r} \qquad ...(iv)$$

5. Electric Field Due to a Group of Charges

The resultant intensity ($\vec{E}$) in such a case is given by

$$\vec{E} = \vec{E_1} + \vec{E_2} + \vec{E_3} + \ldots$$

or
$$\vec{E} = \Sigma \vec{E_n},$$

where $\vec{E_n}$ is the electric field at the point P due to nth point charge, where $n = 1, 2, 3$.

If the charge distribution is continuous then $\vec{E}$ is given by

$$\vec{E} = \int d\vec{E}$$

where $d\vec{E}$ is electric field at the point P due to an infinitesimal element of charge dq at distance r from P.

In S.I. units the magnitudes of $d\vec{E}$ is given by

$$d\vec{E} = \frac{1}{4\pi\varepsilon_0} \cdot \frac{dq}{r^2}$$

6. Electric Field Due to an Infinite Line of Charge

In such a case,

$$E = \frac{2\lambda}{4\pi\varepsilon_0 \cdot r} \text{ or } \frac{\lambda}{2\pi\varepsilon_0 \cdot r}$$

In c.g.s. units the above electric field for air is given by $E = \frac{2\lambda}{r}$ and in a medium of dielectric constant k, it becomes

$$E = \frac{2\lambda}{k \cdot r}$$

7. Electric Field Due to a Finite Line of Charge

In such case, it is given by

$$E = \frac{q}{2\pi\varepsilon_0 \cdot r} \cdot \frac{1}{l}$$

or $$E = \frac{q}{2\pi\varepsilon_0 \cdot r} \quad \left(\text{where } \lambda = \frac{q}{l}\right)$$

λ is the linear charge density.

8. Coulomb's Law

The force of attraction or repulsion between two point charges is directly proportional to the product of the charges and inversely proportional to the square of the distance between them.

$$F = \frac{1}{4\pi\varepsilon_0} \left(\frac{q_1 q_2}{r^2}\right) \text{ newton.}$$

where $\frac{1}{4\pi\varepsilon_0} = 9.0 \times 10^9$ newton-metre^2/coulomb^2.

9. Electric Intensity

At a point is the force experienced by a unit positive charge at that point.

$$\vec{E} = \frac{\vec{F}}{q_0} \text{ Newton/Coulomb.}$$

10. Electric Potential

At a point is the potential energy of unit positive charge placed at that point. It can also be defined as the work done in taking a unit positive charge from infinity to that point.

$$v = \frac{1}{4\pi\varepsilon_0} \left(\frac{q}{r}\right) \text{ volt}$$

Volt. It is the practical unit of potential. The potential difference between two points is one volt if 1 joule of work is done in taking one coulomb from one point to another.

11. Electric Dipole

A pair of equal and opposite point charges separated by a vector distance $2l$ is called an electric dipole.

$$\text{Electric dipole moment} = 2l \times q$$

Its M.K.S. or S.I. unit is Coulomb-metre.

12. Couple on an Electric Dipole in a Uniform Electric Field

$$\tau = PE \sin \theta \text{ newton-metre.}$$

13. Work done in Deflecting a Couple Through θ

$$W = P.E \ (1 - \cos \theta).$$

if $\theta = 90°$, $W = P \times E$, if $\theta = 180°$, $W = 2PE$

14. Electric Intensity due to an Electric Dipole

(i) At a point on its axis, $E = \dfrac{1}{4\pi\varepsilon_0} \cdot \dfrac{2p}{r^2}$.

(ii) At perpendicular bisector, $E = \dfrac{1}{4\pi\varepsilon_0} \times \dfrac{p}{r^2}$.

15. Electric Potential Along axis of Dipole

$$v = \dfrac{1}{4\pi\varepsilon_0} \dfrac{1}{r^2} \text{ volt.}$$

16. Electric Potential on Perpendicular Bisector

Potential on any point on bisector = zero.

17. Equipotential Surface

All the points on such a surface have the same potential. The electric lines of force and the equipotential surface are mutually perpendicular to each other.

18. Electric Field and Potential of a Spherical Shell of Uniform Charge Density

(i) **Outside the Shell**

$$E = \frac{1}{4\pi\varepsilon_0}\left(\frac{q}{r^2}\right) \text{ newton/coulomb}$$

$$v = \frac{1}{4\pi\varepsilon_0}\left(\frac{q}{r}\right) \text{ volt}$$

where r is the distance from centre of shell to the point.

(ii) **On the Surface of Shell**

$E = 0$, v equal to that at the surface.

19. Elementary Charge

It refers to the charge on the object which is always an integral multiple of charge on an electron.

20. Capacity of a Conductor

It is the ratio of charge given to a conductor and potential produced by it.

$$c = \frac{q}{v}$$

The unit of capacity if *farad*.

21. Capacity of Spherical Conductor

$c = 4\pi\varepsilon_0\, a$ farad, where a is the radius in metres.

22. Potential Energy of a Charged Conductor

It is the total work done in charging the conductor to a given potential.

$$E = \frac{1}{2}qv = \frac{1}{2}cv^2 \; \frac{1}{2}\frac{q^2}{c}.$$

Definitions and Formulae in Physics

23. Loss of Energy in Connecting Two Conductors

$$\text{Loss of Energy} = \frac{c_1 c_2 (v_1 - v_2)^2}{c_1 + c_2}$$

24. Capacitor

It refers to an arrangement of conductors whereby the capacity of an insulated charged conductor is considerably increased when the earth connected conductor is brought in its proximity.

25. Factors on which Capacity of a Conductor Depends

(i) $c \infty \dfrac{1}{d}$ (d = distance between two plates)

(ii) $c \infty A$ (A = Area of plates of capacitor

(iii) $c \infty K$ (K = dielectric constant of the medium)

26. Permittivity

$$E_0 = \frac{\sigma}{\varepsilon_0} \text{ and } E = \frac{\sigma}{\varepsilon}$$

where ε is the permittivity of the dielectric and ε_0 is the permittivity of vacuum.

Thus *permittivity* of a medium can be defined as the response of a medium to presence of an electric field.

$$\varepsilon_0 = 8.86 \times 10^{-12} \text{ coulomb}^2/\text{newton-m}^2.$$

27. Dielectric Constant

It refers to the ratio of permittivity of the dielectric and permittivity of vacuum. It is represented by k.

$$k = \frac{\varepsilon}{\varepsilon_0} \text{ or } \varepsilon = k \cdot \varepsilon_0.$$

k is a pure number.

28. Capacity of a Capacitor (c)

$$c = \frac{q}{v_1 - v_2}$$

where q is the charge on plate P_1 and $(v_1 - v_2)$ is the potential difference between plates P_1 and P_2.

29. Capacity of a Parallel Plate Air Capacitor

$$c_0 = \left(\frac{\varepsilon_0 A}{d}\right) \text{ farad}$$

30. Capacity of a Parallel Plate Capacitor Filled Completely with Dielectric

$$c = c_0 \times k = \frac{\varepsilon_0 \cdot k \cdot A}{d} = \frac{\varepsilon \cdot A}{d}$$

31. Parallel Plate Capacitor Partially Filled with Dielectric

$$c' = \frac{\varepsilon_0 \cdot A}{\left[d - t\left(1 - \frac{1}{k}\right)\right]}$$

32. Capacity of a Spherical Capacitor

$$Q = 4\pi\varepsilon_0 \frac{k \cdot r_1 \cdot r_2}{r_2 - r_1} \text{ farad}$$

where r_1 and r_2 are radii of inner and outer spheres.

33. Capacitor in Series

$$\frac{1}{c} = \frac{1}{c_1} + \frac{1}{c_2} + \frac{1}{c_3} + \ldots$$

34. Capacitor in Parallels

$$c = c_1 + c_2 + c_3 + \ldots$$

35. Vande Graaff Generator

It is an electrostatic generator capable of producing voltage of the order of 8 million volts. It was designed in 1931 in U.S.A. by Vande Graaff.

36. Mev and Gev

$$1\ Mev = 10^6\ ev;\ 1\ Gev = 10^9\ ev.$$

Mev is million electron volt and Gev is giga electron volt.

Electric current (i)

$i = \dfrac{\Delta q}{\Delta t}$, 1 ampere = 1 coulomb per second.

(i) Current in a conductor is due to drift of electrons.

(ii) Current in an electrolyte is due to +tive or —tive ions.

(iii) Current in a gas is due to drift of both kinds of ions.

(iv) Current in a semi-conductor is due to drift of free electrons and holes.

30

Ohm's Law and Kirchhoff's Law

1. Electric Current

The flow of charge is called current. The free charges are negative electrons and they flow from a higher potential to a lower potential. The practical unit of electric current is *Ampere*. An ampere is that much current which flow in a conductor when one coulomb of electricity passes through any cross section of it in one second. A current of one ampere means 6.2×10^8 electrons flowing per second through any section of the wire.

When no current is applied across a conductor, the free charges are in thermal equilibrium with rest of the conductor and are in random motion and in such a case there is no current in the conductor.

If a net charge 'q' passes through any cross-sectional area of conductor in time 't', the electric current 'i' is given by

$$i = \frac{q}{t}$$

Definitions and Formulae in Physics

If the magnitude of current is a function of time the electric current is given by

$$i = \frac{dq}{dt}.$$

2. Current Density

The current 'i' is characteristic of a particular conductor and is a macroscopic property. A related microscopic quantity is current density J. It is a vector quantity and is the characteristic of a point inside the conductor and not of the conductor as a whole. If the current is distributed uniformly across a conductor then the magnitude of current density is equal to current through infinitesimal area at that point, i.e.,

$$J = i A$$

the area being normal so the direction of flow of current.

When the plane of small area A makes an angle θ with the direction of current, then

$$J = \frac{i}{A \cos \theta} \text{ or } i = J A \cos \theta = J \cdot A$$

The unit of current density is amp/metre^2.

3. Drift Velocity

When no potential difference is applied across a conductor, the electrons are in random motion. The average velocity of electrons is zero. Thus the motion of electrons does not constitute any transport of charge in any direction. The current in the conductor is zero. When a potential difference is maintained across a conductor, the electrons gain some average velocity in the direction of positive potential. This average velocity is superimposed over the random velocity and is known as drift velocity.

Let us consider the simple charge distribution of electrons each of charge e and N be the number of electrons per unit volume then

$$P = N\,e \text{ and } J = N\,e\,v_d$$

or $\qquad v_d = Jj/n\,e$ or $v_d = i/N\,A\,e$

where v_d is the drift velocity.

Potential Difference. V between two points in a conductor or is measured by the work W done in transferring unit charge from one point to the another.

$\therefore \qquad V$ (Potential difference).

$$= \frac{W(\text{work done transfer charge})}{q(\text{charge transferred})}$$

The potential difference will be 1 volt if joule of work is required to transfer 1 coulomb of charge from one point to the other

$\therefore \qquad V(\text{volt}) = \dfrac{W(\text{joule})}{q(\text{coulomb})}.$

Electromotive Force E. An agent such as a battery or generator has an electromotive force (e.m.f.) if it does work on the charge moving through it. The e.m.f. is measured by the p.d. between the terminals when the battery or generator is not delivering current. Its unit is MKS system is 1 volt (1 joule per coulomb).

Definitions and Formulae in Physics

Resistance (R) of a conductor is the property which depends on its dimensions, material and temperature and which determines the current produced in it by a given potential difference across its ends. Its unit is *ohm*.

One ohm is the resistance of a conductor in which a current of 1 ampere exists when the potential difference between its ends is 1 volt.

$$R \text{ (resistance)} = \frac{V \text{ (Potential difference)}}{1 \text{ (Current)}}$$

Thus R (ohm) $= \dfrac{V \text{(volt)}}{1 \text{ (ampere)}}$

The above relation is known as *Ohm's law*. It states that the current I in a metallic conductor at a constant temperature is equal to the potential difference V across the ends of the conductor divided by the resistance R of the conductor.

The terminal voltage of a battery when it delivers a current I is equal to the total e.m.f. E minus the potential-drop (voltage drop) in its internal resistance r. Thus

(i) When delivering current,

Terminal voltage = e.m.f. — potential drop in the internal resistance = $E - Ir$

(ii) When receiving current,

Terminal voltage = e.m.f. + voltage drop in internal resistance

$$= E + Ir$$

(iii) When no current exists.

(open circuit)

Terminal voltage = E of the battery.

4. Resistivity of Specific Resistance

The resistance of a conductor of length l metre and cross sectional area A metre^2 is

$$R = K \frac{l}{A} \text{ in ohms.}$$

the constant K is called *resistivity* or *specific resistance* of the material of the conductor. Its unit is *ohm-metre*.

Electrical Conductivity. *The electrical conductivity of a material is defined as the reciprocal of the resistivity i.e.,*

$$\text{Electrical conductivity } \sigma = \frac{1}{\text{resistivity } \rho}$$

$\therefore \qquad \rho = l/RA$

Its unit is (ohm-metre)$^{-1}$ or ohm/metre.

Variation of resistance with temperature. The resistance of a conductor varies with temperature. If R_t be the resistance of a conductor at $t°C$ and R_0 its resistance at $0°C$, then

$$R_t = R_0 (1 + \alpha t)$$

where α is called temperature coefficient of resistance

$$\alpha = \frac{R_t - R_C}{R_0 t}.$$

The temperature coefficient or resistance may be defined as the fractional increase in the resistance of the conductor per unit increase in its temperature.

5. Series and Parallel Circuits

For series circuit

(i) The current is the same in every part of the circuit.

(ii) The combined resistance R of the circuit is equal to the sum of the separate resistance :

$$R = R_1 + R_2 + R_3 + \ldots..$$

(iii) The potential difference V across the combination is equal to the sum of potential differences across the separate resistances.

For parallel circuit

(i) The sum of the currents in the branches is equal to the main current of the circuit.

(ii) The equivalent resistances R of the parallel combination of resistances $R_1, R_2, R_3\ldots$ is given by the relation

$$\frac{1}{R} = \frac{1}{R_1} + \frac{1}{R_2} + \frac{1}{R_3} + \ldots.$$

(iii) The potential difference across several conductors in parallel is the same as across each of the conductors.

6. Grouping of Cells

Series arrangement. Fig shows n cells, each of emf E and internal resistance r ohm connected in series. A resistance R ohm is connected in the external circuit. The total emf in this case

$$= nE \text{ volts}$$

Total internal resistance

$$= nr \text{ ohms}$$

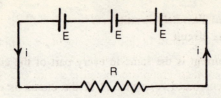

Total resistance of the circuit

$$= (R + nr) \text{ ohms}$$

$$\therefore \quad \text{Current } i \frac{nE}{(R + nr)} \text{ amperes}$$

(i) If $R \gg nr$, then $i = nE/R$

(ii) If $R \ll nr$, then $i = E/r$

Parallel arrangement. Fig below shows n cells, each of emf E and internal resistance r ohm connected in parallel. A resistance R ohm is connected in external circuit.

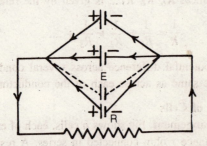

The effective internal resistance $= r/n$ ohm

Total resistance of the circuit $= (R + r/n)$

$$\therefore \quad \text{current } i = \frac{E}{[R + (r/n)]} \text{ amp.}$$

(i) If R is small then $i = nE/r$.

(ii) If R is large, then $i = E/R$, same as that of single cell.

Mixed arrangement. In this arrangement the total number of cells N are divided into m groups. Each group has n cells. The n cells in each group are joined in series and then all the m groups are arranged in parallel. In this case

$$\text{current } i = \frac{NE}{mR + nr}.$$

7. Kirchhoff's Laws

(i) At any point in a circuit the sum of the currents flowing towards the point is equal to the sum of the currents flowing away from the point.

(ii) In any closed circuit the sum of the IR drops around any path is equal to the e.m.f.s. impressed on that path.

Let us apply Kirchhoff's second law to Fig below :

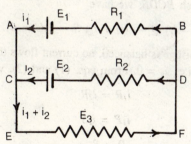

For the mesh $ACDBA$,

$$i_1 R_1 - i_2 R_2 = E_1 - E_2$$

For the mesh $EFDCE$,

$$i_2 R_2 + (i_1 + i_2) R_3 = E_2$$

For the mesh $EFBAE$,

$$i_1R_1 + (i_1 + i_2) R_3 = E_1.$$

8. Condition of Balance in Wheatstone's Bridge

The wheatstone's bridge is shown in Fig. below. In order to consider the condition of balance in this bridge, we apply Kirchhoff's law to different meshes.

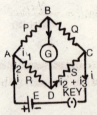

Applying Kircchoff's second law to mesh *ABDA*, we have

$$i_1P + i_gG - i_2R = 0 \qquad \ldots(1)$$

For the mesh *BCDB*, we have

$$(i_1 - i_g) Q - (i_2 + i_g) S - i_gG = 0 . \qquad \ldots(2)$$

When the bridge is balanced, no current flows through galvanometer G i.e., $i_g = 0$. From eqs. (1) and (2), we have

$$i_1P = i_2R \qquad \ldots(3)$$

$$i_1P = i_2S \qquad \ldots(4)$$

From eqs. (3) and (4) $\dfrac{P}{Q} = \dfrac{R}{S}$

This is the condition for the balance of wheat stone's bridge.

31

Heating Effect of Current

1. Electric Energy and Heat

When a current i amperes is passed for t seconds between two points having a potential difference of V volts, the workdone W is given by

$$W = V \, i \, j \text{ joules}$$

Thus the electric energy used in t seconds $= V \, i \, t$ joules.

According to Ohm's law, $V = i \, R$ ...(1)

∴ Electric energy $(iR) \, i \, t = i^2 R \, t$ ...(2)

or Electric energy $= V \left(\dfrac{V}{R}\right) t = \dfrac{V^2}{R} t$ ...(3)

The *MKS* unit of electric energy is joule but it is generally expressed in kilo watt-hour (kwh).

$$1 \text{ kilo watt-hour} = 1000 \times 60 \times 60 \text{ J}$$

$$= 36 \times 10^5 \text{ joule}.$$

The electrical appliances are connected in parallel and are electrical energy consumed is measured in k.w.h.

$$\text{No. of units} = \frac{\text{watt} \times \text{hour}}{1000}$$

Electric Power. The energy liberated per second in an electric device is called electric power P

$$P = \frac{Vit}{t} = Vi = i^2 R$$

$$= \frac{V^2}{R} \left(\text{By Ohm's law } i = \frac{V}{R} \right) \qquad ...(4)$$

The unit of power is watt.

Joule's Law of Heating by Electric Current. Heat produced = $\dfrac{\text{current}^2 \times \text{resistance} \times \text{time}}{4.2}$

or
$$H = \frac{C^2 Rt}{4.2} = 0.24 \, C^2 Rt$$

where H is in calories
C is in amperes
R is in ohms
t is in sec.

2. Thermoelectric Effects and their Applications

The production of heat by passage of an electric current is the most familiar example of the irreversible thermoelectric phenomenon.

Some thermoelectric phenomena are found to exist which are reversible. Such a reversible phenomenon occurs at the junction of two dissimilar conductors maintained at different temperatures. This effect is called *Seebeck effect*.

The term thermoelectric effect refers to such reversible phenomenon called *Seebeck effect*.

The wires of two different metals say copper and iron, are joined together at the end A and B (See Fig.).

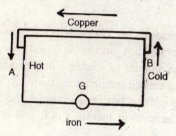

On heating junction A, current in the direction iron to copper at the cold junction and copper to iron at the hot function will begin to flow.

The continuous flow of current indicates that there is e.m.f produced in the circuit. This is called thermoelectric e.m.f.

The combination of two metals is called thermocouple.

This generation of e.m.f in a thermocouple when one of the junctions is heated keeping the other cold is called Seeback effect.

The magnitude of e.m.f. produced depends upon the nature of metals in contact and the difference in temperature of the two junctions.

It is maximum for bismuth-antimony thermocouple.

From the following series if any metal is chosen for use in a thermocouple the current will flow from the metal earlier in series to the metal later in the series across the cold junction.

Sb, Fe, W, Zn, Ag, Au, Cr, Sn, Ph, Hg, Mn, Cu, Pt, Co, Ni, Bi.

The more separated are the metals in this series more will be the e.m.f. produced.

3. Effect of Temperature on Thermoelectric e.m.f.

As the temperature of hot junction is raised (keeping cold junction at 0°C), the amount of thermo e.m.f. produced goes on increasing upto a certain temperature (upto 270°C in case of Cu- Fe thermocouple) on heating junction A (hot junction) further there is a decrease in thermo e.m.f. and it becomes

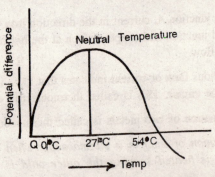

zero at certain temperature (at 540°C in Cu-Fe thermocouple). On heating beyond this temperature the e.m.f. is produced in the reverse direction.

The temperature at which thermoelectric current is maximum is called the *neutral temperature*.

The temperature at which the direction of current is reversed is called the *temperature of inversion.*

The temperature of inversion is as much above the neutral temperatures as the neutral temperature is above the temperature of cold junction.

4. Applications
(1) **Thermoelectric thermometers**

Merits

(a) These are cheap and easy to construct.

(b) They have a low thermal capacity.

(c) They can be used to measure temperature at a specific point.

(d) They can be used over a wide range. Approx. 200° to 1600°C.

(e) For temperatures above 1000°C they are the only sensitive and convenient to use.

Demerits

(i) There is no formula which relates the temperature difference between the two junctions and the e.m.f.

(ii) Below 1000°C, they are inferior in accuracy to platinum resistance thermometers.

(iii) Because of difficulty of maintaining the cold junction at 0°C a correction has to be applied.

(2) **Thermopile.** It is sensitive instrument and is used to detect the presence of heat radiations.

(3) **Boy's radio micrometer.** It was designed by Prof. C.B. Boys in 1888. It is used to measure very small amounts of thermal radiation.

(4) **The thermoelectric pyrometer.** It is used to measure high temperatures such as those of furnaces.

(5) **Duddel's thermo-galvanometer.** It is a very sensitive ammeter and is used equally for measuring direct and alternating currents.

(6) **Peltier effect.** The phenomenon according to which heat energy is absorbed or evolved at a junction of a thermocouple when an electric current is made to pass through i is called *Peltier effect*.

Peltier coefficient. The amount of heat energy (in joules) evolved or absorbed at a junction of two dissimilar metals when 1 ampere of current flows across the junction for one second is called peltier coefficient (π).

The Peltier coefficient is different for different pairs of metals, its value varies with the absolute temperature of the junction.

The Peltier coefficient at a junction expressed in joules, is numerically equal to the contact potential difference (in volts) at the junction.

(7) **Thomson effect.** The phenomenon of absorption or evolution of heat along an unequally heated conductor when an electric current is made to pass through it is called *Thomson Effect*.

Cu, Sb, Ag, Cd, etc. show positive Thomson effect.

Ee, Ni, Co, Bi, etc., show negative Thomson effect.

Please note that while Thomson effect deals with the heat energy evolved or absorbed along the length of an unequally heated conductor, the Peltier effect deals with the heat energy evolved or absorved at the junction of a thermocouple.

Thomson coefficient. The amount of heat energy (in joules) evolved or absorbed between the two points of a conductor, differing in temperature by 1°C when 1 ampere of current flows through it for 1 sec is called *Thomson Coefficient* (σ).

The value of Thomson coefficient is different for different conductors and for a particular conductor it varies with temperature.

The Thomson coefficient (expressed in joules) of a conductor is numerically equal to the potential difference in volts developed between the points of the conductor differing in temperature by 1°C. Thomson effect is also called specific heat of electricity.

32

Magnetic Effects of Current

1. Magnetic Field of a Current

Consider that a current I ampere exists in a wire element of length Δl metre. Then the *magnetic induction* or *magnetic field* ΔB in weber/metre^2 at point O due to the current element is given by

$$\Delta B = K \frac{I \Delta l \sin \theta}{r^2}$$

where r is metre is the distance from Δl to O; and θ is the angle between r and the current element in Δl. K is a constant

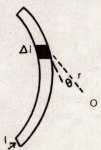

Definitions and Formulae in Physics

and its value for vacuum in MKS system 10^{-7} weber per ampere-metre. K has practically the same value for air and for non-ferromagnetic substances.

2. Magnetic Field of a Circular Coil

The magnetic field at the centre of a circular coil of radius r metre due to a current I ampere in it is given by the relation

$$B \text{ (weber/metre}^2) = K \frac{2\pi n I}{r}$$

where $K = 10^7$ weber per ampere-metre and n is the number of turns in the coil.

3. Magnetic Field Due to a Long Straight Wire

$$B = R \frac{2I}{r}$$

where B is in weber/metre^2
K is 10^{-7} weber per ampere-metre
I is current in ampere
r metre is the perpendicular distance of the point from the wire.

Magnetic field of a Solenoid. Consider a long solenoid having n turns of wire and length l metres, and carrying a current I amperes. The magnetic field B at any point in the *interior* is

$$B = R \frac{4\pi n I}{l}$$

where C is in weber/metre^2 and
$K = 10^{-7}$ weber per ampere-metre.

If the solenoid is wound in the form of a *toroid* (ring) the same equation holds true if l = means circumference of the toroid. The magnetic field is entirely *within* the toroid.

4. Force on a Current Carrying Conductor in a Magnetic Field

For a straight conductor of length l metre in a uniform field of B weber/metre^2, the force on the conductor is

$$F \text{ (newton)} = Bl\,I \sin \theta$$

where I ampere is the current in the conductor and θ is the angle between the field and the conductor.

The force is perpendicular both to the current and the field.

5. Fleming's Left Hand Rule

"Extend thumb, forefinger and middle finger of the left hand at right angle to one another. If the fore-finger points in the direction of the field and the middle finger points in the direction of the current, then the thumb will indicate the direction of the motion of the conductor."

6. Charged Particle in Uniform Constant Magnetic Field

When a test charge q moves in a magnetic field of induction B with a velocity v, then it experiences a side way deflecting force F (see Fig. below), given by

$$F = a\,(v \times B)$$

or $\qquad F = q\,v\,B \sin \theta \qquad \ldots(1)$

Thus the force F is always perpendicular to v, and B.

It is clear from eq. (1) that

(i) If the particle is at rest in side the field, no force will act on it ($v = 0$) and hence the particle remain at rest.

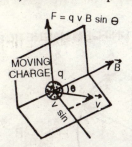

(ii) If the particle is moving parallel to magnetic field ($\theta = 0$), no force acts on it. Thus a charged particle initially moving parallel to magnetic field will continue to move with initial constant speed on parallel path.

(iii) If the particle is moving perpendicular to magnetic field, it experiences a maximum force.

Thus, we can say that force depends v, B and the sine of angle θ between them.

Now we shall describe the motion of charged particle in a magnetic field. Consider the case of a charged particle having charge q enters in a magnetic field at right angle to the direction of field as shown in Fig. given below.

As the force due to magnetic field on the charged particle is perpendicular to the direction of motion of particle and direction of field, hence no work is done by magnetic field on the charged particle. This shows that the particle does not gain kinetic energy. Thus the velocity of particle is changed. As v and B are constant, F is also constant and is maximum which is represented by F_m, the force provides the necessary centripetal force. Hence the path traversed by the particle will be circular.

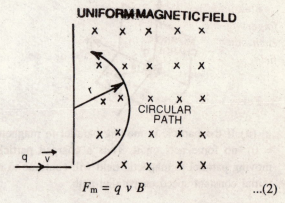

Now
$$F_m = q\,v\,B \qquad ...(2)$$

Let a be the acceleration produced by this force, then

$$F_m = ma = m\frac{v^2}{r} = q\,v\,B \qquad ...(3)$$

From eq. (3) $\quad mv = qBr$ or $r = mv/qB \qquad ...(4)$

The angular velocity ω is given by

$$\omega = \frac{v}{r} = \frac{qB}{m} \text{ or } f = \frac{\omega}{2\pi} = \frac{qB}{2\pi m} \qquad ...(5)$$

The frequency is independent of v. This frequency is known as cyclotron frequency.

Here it should be remembered (eq. 4), that faster particles move in bigger circle and slower particles move in smaller circle such that time period T is the same.

When a charged particle moves in electric as well as magnetic field, then the force is given by

$$F = qE + q(v \times B) \qquad ...(6)$$

This force is known as Lorentz force.

Fleming's left hand rule. *If the thumb and the first two fingers of the left hand are held mutually perpendicular to each other, and the first finger is pointed in the direction of field, the second finger in the direction of current then thumb gives the direction of force.*

7. Force on a Current Carrying Conductor

Consider the case of a current carrying conductor placed in magnetic field as shown in Fig. given below.

We know that current is an assembly of moving charges, therefore, the magnetic field will exert a side way force on the conductor carrying a current. If the charge is infinite small, then

$$dF = dq(v \times B) \qquad ...(1)$$

where v is displacement of charge per unit time.

∴ $$dF = dq\left(\frac{dl}{dt} \times B\right) = \frac{dq}{dt}(dl \times B)$$

∴ $$v = dl/dt \text{ or } dF = i(dl \times B) \qquad ...(2)$$

When B is uniform over the length of wire, then integrating eq. (2), we have

$$F = i\,(l \times B) \qquad ...(3)$$

The force is perpendicular to both l and B.

8. Force Between Two Parallel Conductors

Fig. shows a portion of two long straight parallel conductors a and b separated at a distance d and carrying current i_1 and i_2 respectively. Here each conductor lies in the magnetic field set up by other conductor and hence each will experience a force.

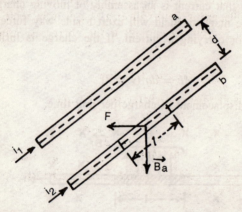

Let B_a be the field of induction produced by conductor a carrying a current i_1, near all points of conductor b. The magnitude of B_a is given by

$$B_a = \frac{\mu_0}{4\pi} \cdot \frac{2i_1}{d} \qquad ...(1)$$

The force on a length l of this conductor is

Definitions and Formulae in Physics

$$F = i_2 B_a l$$

$$= \frac{\mu_0}{4\pi} \cdot l \cdot \frac{2i_1 i_2}{d} \qquad \ldots(2)$$

The force per unit length is given by

$$\frac{F}{l} = \frac{\mu_0}{4\pi} \frac{2i_1 i_2}{d} \qquad \ldots(3)$$

The direction of B_a at wire b will be downward and direction of F will be towards left.

Similarly we can start with wire b, compute its field of induction B_b at wire a and then find the force on wire a. The direction of force will be towards right. The two forces will be equal and opposite. Hence the conductors attract each other.

If the direction of either current is reversed, the force will be in opposite directions. Thus, the parallel conductors carrying current in opposite directions repel each other.

33

Galvanometer and Measuring Instruments

1. Ammeter

A sensitive galvanometer can be converted into an ammeter of any required range. The ammeter should be capable of recording the correct value of the current flowing in the circuit without affecting the circuit current. Hence it should have a low resistance. So a shunt of low resistance is connected in parallel to the galvanometer resistance to convert a galvanometer into ammeter as shown in Fig. below :

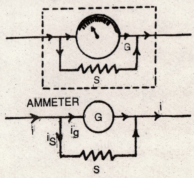

To convert a galvanometer which gives full scale deflection for a current i_g, so that it may be used to read a current i, the value of the shunt required is given by

$$S = \frac{i_g G}{i - i_g}$$

where G = galvanometer resistance.

2. Voltmeter

A sensitive galvanometer can be converted into a voltmeter of any required range. The voltmeter should be such that when connected between two points to measure potential difference, a small current should pass through it. Hence a galvanometer may be used as a voltmeter of desired range by connecting suitable high resistance in series with it as shown in Fig. below :

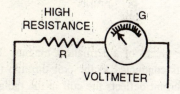

The value of resistance R connected in series is given by

$$R = \frac{V}{i_g} - G$$

where, V = maximum voltage to be measure
i_g = current through the galvanometer giving full scale deflection.

3. Moving Coil Galvanometer

Moving coil galvanometer is used for the measurement of current. This is based on the principle that when a current carrying conductor is placed in the magnetic field it experiences a force which is given by Fleming's left hand rule. The moving coil galvanometer is shown in Fig. below. It essentially consists of a rectangular coil *PQRS* of large number of turns of fine insulated copper wire wound over a nonmagnetic metallic frame which is suspended by a phosphor bronze wire

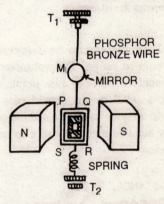

between the pole pieces of a permanent magnet. The current to be measured is conducted to the coil through the suspension wire. The current deflects the coil in the radial magnetic field between the soft iron cylinder E and the concave pole pieces. The amount of deflection serves as a measure of current. The deflection is measured with the help of a mirror M attached to phosphor bronze wire using lamp and scale arrangement.

The current i is directly proportional to deflection produced θ i.e.,

$$i \propto \theta \quad i = K\theta$$

Definitions and Formulae in Physics

where K is a constant of the galvanometer and is known as galvanometer constant.

$$K = \frac{C}{NAB}$$

where C = elastic torisonal constant of the suspension.
N = number of turns in the coil.
A = area per turn of the coil.
B = magnetic induction of radial magnetic field.

34

Faraday's Laws of Electrolysis

1. Faraday's Laws of Electrolysis

(i) The mass of a substance liberated or deposited at an electrode is proportional to the quantity of electricity that has passed through electrolyte.

(ii) The masses of different substances liberated or deposited in electrolysis by the same quantity of electricity are proportional to their equivalent weights.

Mathematically Faraday's laws are expressed as

m = mass deposited or liberated (gm)
$m = ZCt$ C = current (ampere)
t = time (second)

Z is called the *Electro-chemical equivalent* and is defined as the mass deposited or liberated when 1 coulomb of electricity passes through the electrolyte. Its unit is "gm/coulomb".

For copper $Z = 0.00033$ gm/coulomb

For silver $Z = 0.00118$ gm/coulomb

Definitions and Formulae in Physics

One Faraday (= 96,500 coulombs) is the quantity of electricity required to deposit 1 gm equivalent weight of any substance in electrolysis. If m gm of a substance is liberated in electrolysis, then

m = gm equivalent weight × number of faradays transferred.

35

Electromagnetic Induction

1. Electromagnetic Induction

An e.m.f. is induced in a conductor whenever there is a change in the magnetic flux lined with the conductor. This phenomenon is called *electromagnetic induction*. If the circuit is complete, an induced current will result.

Faraday's Law of Electromagnetic Induction state that the e.m.f. induced in a conductor is proportional to the time rate of change of flux through it.

$$|e| = \frac{d\varphi}{dt}$$

Lenz's law states that the direction of the induced current is such that its own magnetic field opposes the change in flux responsible or inducing the current.

Combining the two laws we have

$$E = -n\frac{d\varphi}{dt}$$

where n is the number of turns in the coil and $\dfrac{d\varphi}{dt}$ is the rate of change of flux for each turn. The minus sign indicates that the induced e.m.f. opposes the cause which produces it.

The e.m.f. E will be in volts if $\dfrac{d\varphi}{dt}$ is in weber per second.

2. EMF Induced in Moving Conductor

The e.m.f. E induced in a straight conductor of length l metre moving with velocity v metre/sec perpendicular to a magnetic field of magnitude B weber/metre^2 is given by the following relation :

$$E = Blv$$
(volt)

Induced charge. When the magnetic flux φ through a closed circuit of known resistance R changes, the quantity of electric charge which flows around the circuit is given by

$$\text{Induced charge} = \dfrac{\text{Change of magnetic flux}}{\text{resistance}}.$$

3. Induction Due to the Motion of a Straight Rod

Consider the case of a conductor AB moving with velocity v on U shaped conductor towards right as shown in Fig. The U shaped conductor is situated in uniform magnetic field pointing in the plane of the paper. The moving conductor causes a change in the area of the circuit. Let the area of the circuit changes from $ABCD$ to $A'B'CD$ in time dt. This causes an increase in flux given by

$$d\varphi = B \times (\text{change in area})$$

or $\qquad d\varphi = B \, (\text{area } A'B'CD - \text{area } ABCD)$

$$= B \text{ area } (ABB'A') = B\, l\, v\, dt$$

The magnitude of induced e.m.f. in the circuit is given by

$$e = \frac{d\varphi}{dt} = B\, l\, v$$

The direction of induced e.m.f. is given by Lenz's law. Thus is shown in fig.

4. Mutual Inductance

Consider two coils P and S placed near to each other as shown in Fig. When current passing through a coil increases or decreases, the magnetic flux linked with the other coil also changes and an induced e.m.f. is developed in it. This phenomenon is known as *mutual induction*. This coil in which current is passed is known as primary and the other in which e.m.f. is developed is called as secondary.

Let the current through the primary coil at any instant be i_1. Then the magnetic flux φ_2 at any point in the secondary will be proportional to i_1, i.e.,

$$\varphi_2 \propto i_1$$

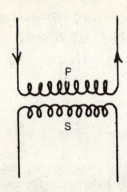

Therefore, the induced e.m.f. in secondary when i_1 changes is given by

$$e = -\frac{d\varphi_2}{dt} \text{ i.e., } e \infty -\frac{di_1}{dt}$$

$$\therefore \quad e = -M\frac{di_1}{dt}$$

where M is a constant of proportionality and is known as mutual inductance of two coils. *It is defined as the e.m.f. induced in the secondary coil by unit rate of change of current in the primary.* The unit of mutual inductance is henry. The mutual inductance of a coil can be increased by wounding it over an iron core. The mutual inductance now becomes μM where μ is permeability of iron.

Mutual inductance of a solenoid

$$M = \mu_0 \, A \, N_1 \, (N_2/l).$$

5. Self-Inductance

When the current flows in a coil, it gives rise to a magnetic flux through the coil itself. When the strength of current changes, the flux also changes and an e.m.f. is induced in the coil. This e.m.f. is called self-induced e.m.f. and the phenomenon is known as self-induction.

The flux through the coil is proportional to current through it, i.e.,

$$\varphi \propto i$$

$$\therefore \quad e = -\frac{d\varphi}{dt} \text{ i.e., } e \propto \frac{di}{dt}$$

where L is constant of proportionality and is called as self-inductance or simply inductance. The unit of inductance is also henry. If the coil is wound over iron core, the inductance is increased by a factor μ (permeability or iron).

Self-inductance of a solenoid.

$$L = \mu_0 N^2 (A/l).$$

36

Cathode Rays and Bohr's Model of Atom

1. Cathode Rays

When a discharge tube is exhausted to a low pressure (10^{-2} mm to 10^{-3} mm), the positive column disappears and Crooke's dark space fills the whole tube. At this stage the walls of the glass tube fluoresces with green colour. This is the stage when bluish steamers travel across the tube from the cathode to anode. These steamers are called cathode rays. Cathode rays are independent of the nature of the gas or the electrodes employed in the discharge tube. In fact, cathode rays are beams of electrons which are common to all elements. When these rays pass through the gas, fresh ions are produced by collision and the discharge is maintained.

For cathode rays :

$e = 1.6 \times 10^{-19}$ coulomb

$m = 9.1 \times 10^{-31}$ kg

$e/m = 1.76 \times 10^{11}$ coulomb/kg.

2. Properties of Cathode Rays

Following are the properties of cathode rays

(i) They travel in straight lines and cast shadows of objects placed in their path.

(ii) They can transfer their energy so as to set a small paddle wheel into rotation.

(iii) They produce heat when allowed to fall on matter.

(iv) They produce flourescence on some substances on which they strike.

(v) Cathode rays can penetrate through small thickness of matter.

(vi) They carry negative charge.

(vii) They affect a photographic plate.

(viii) They ionise the gas through which they pass and makes it conducting.

(ix) They are deflected by a magnetic field.

(x) They are also deflected by electric field.

(xi) When cathode rays strike a solid substance of large atomic weight they produce X-rays.

3. Cathode Rays Oscillograph

Cathode rays oscillograph (C.R.O.) is an instrument which demonstrates the rapid variations of electric current and potentials on a fluorescent screen. This utilises the positive rays for determining unknown frequencies. It is used in radar, television, electric cardiography etc. It consists of the following three main parts :

(i) *Electron gun.* The function of electron gun is to produce, focussing and to obtain a fine pencil of electronic beam.

(ii) *Deflecting system.* The electronic beam is deflected in magnetic and electric fields.

(iii) *Fluorescent screen.* The fast moving electrons strike the screen coated with fluorescent material (like zinc sulphide) and produce visual sensation.

4. Bohr's Theory of Hydrogen Atom
Main assumptions

(i) The electron can exist in certain orbits without radiating energy, even though the radiation would be expected because of the centripetal acceleration on the electron.

(ii) Only those orbits are allowed for which the angular momentum (Iv) is an integral multiple of $h/2\pi$ where h is the Planck's constant. That is

$$I\omega = \frac{nh}{2\pi}.$$

The integer n is called the quantum number.

(iii) If the electron goes from an orbit of energy W_1 to another orbit of lower energy W, then a photon of energy hf is radiated such that

$$hf = W_1 - W$$

Energy of hydrogen atom in the nth energy state is

$$W_n = -K^2 \frac{2\pi^2 me^4}{n^2 h^2}$$

where W_n is in joules, $K = 9 \times 10^9$ nt m^3/coul^2, m is mass of the electron in kg, e is electronic charge in coulombs, n is the quantum number and h is Planck's constant = 6.62×10^{34} joule/sec

The radius r of the nth orbit is given by

$$r = \frac{1}{K} \frac{n^2 h^2}{4n^2 me^2} \text{ (in MKS system)}.$$

If the wavelength of the radiation emitted from the hydrogen atom is λ metres, then we have the relation

$$\frac{1}{\lambda} = \frac{K^2 2\pi^2 m e^4}{h^3 c} \left(\frac{1}{n_1^2} - \frac{1}{n_2^2} \right)$$

indicating that the electron has jumped from the nth orbit to n_{1th} orbit of lower energy.

For Balmer series, we have

$$\frac{1}{\lambda} = R \left(\frac{1}{2^2} - \frac{1}{n^2} \right) \text{ where } n = 3, 4, 5 \text{ etc.}$$

The constant $R = 1.10 \times 10^7$ metre^{-1} and λ is in metres. The constant R is called *Rydberg constant*.

Based on these assumptions some important formulae have been obtained which are used to explain spectral series of atomic hydrogen.

(i) The radius of the nth orbit of Bohr's atom of atomic number Z is given by

$$r_n = \frac{n^2 h^2}{4p^2 m Z e^2} \text{ in MKS units}.$$

Definitions and Formulae in Physics

$$r_n = \frac{(4\pi\varepsilon_0)\, n^2 h^2}{4\pi^2 m\, Ze^2}$$

where e and m are respectively the charge and the mass of the electron.

(ii) The energy of the electron in the nth orbit is given by

$$E_n = \frac{2\pi^2 Z^2 m e^4}{n^2 h^2}$$

$$E_n = -\frac{1}{(4\pi\varepsilon_0)^2} \cdot \frac{2\pi^2 m\, Z^1 e^4}{n^2 h^2}, \text{ in MKS system}$$

$$= -\frac{h\, RcZ^2}{n^2}$$

where
$$R = \frac{2\pi^2\, m e^4}{c h^3}$$

$$R = \frac{1}{(4\pi\varepsilon_0)} \cdot \frac{2\pi^2 m e^4}{c h^3} \text{ in MKS system}$$

is known as Rydberg constant.

(iii) When the electron jumps from an outer orbit with energy E_i and quantum number n_f to an inner orbit with energy E_f and quantum number n_f, the frequency of light emitted is given by

$$v = \frac{E_i - E_f}{h}$$

$$= -\frac{Rc\, Z^2}{n_i^2} - \left(-\frac{cRZ^2}{n_f^2}\right)$$

$$= Rc\, Z^2 \left(\frac{1}{n_i^2} - \frac{1}{n_f^2}\right)$$

The wave number (i.e., the reciprocal of wavelength) of the emitted radiation is given by

$$\bar{v} = \frac{1}{\lambda} = \frac{v}{c}$$

$$= RZ^2 \left(\frac{1}{n_i^2} - \frac{1}{n_f^2}\right)$$

For hydrogen atom $Z = 1$, therefore, we have

$$\bar{v} = R \left(\frac{1}{n_i^2} - \frac{1}{n_f^2}\right)$$

where
$$R = \frac{13.6\, eV}{hc}$$

$$= 1.098 \times 10^7 \text{ metre}^{-1}.$$

5. Series in the Hydrogen Spectrum

For different fixed values of n_f, and taking n_i, as running integer, we obtain the different series of hydrogen spectrum. For

(i) **Lyman series** $n_f = 1$, $n_i = 2, 3, 4,...$

(ii) **Balmer series** $n_f = 2$, $n_i = 3, 4, 5,...$

(iii) **Paschen series** $n_f = 3$, $n_i = 4, 5, 6,...$

(iv) **Brackett series** $n_f = 4$, $n_i = 5, 6, 7,...$

(v) **Pfund series** $n_f = 5$, $n_i = 6, 7, 8,...$

6. Electron Volt

It is the kinetic energy gained by an electron when it is accelerated through a potential difference of one volt.

Definitions and Formulae in Physics

1 electron volt (1 eV)

$= 4.8 \times 10^{-10}$ e.s.u. of charge

$\times \dfrac{1}{300}$ e.s.u. of potential

$= 1.6 \times 10^{-12}$ erg

Also 1 $eV = 1.6 \times 10^{-19}$ coulomb $\times$ 1 volt

$= 1.6 \times 10^{-19}$ joule.

37

Diodes and Triodes

1. Thermionic Emission
It is the process of emission of electrons from an emitter by supplying *heat energy*.

If a charged particle is produced as a result of heating a substance, it is known as *thermionic* and the process is known as *thermionic emission*.

2. Richardson Equation
It is an equation that gives a relation between the saturated thermionic current (I) and the absolute temperature (T) to which a metal is heated. It is

$$I = AT^2 e^{-\varphi/KT},$$

where A = constant depending upon tube
K = Boltzmann constant
φ = work function.

3. Thermionic Tube
It is an evacuated glass tube having two or more electrodes, cathode, one of the electrodes, is taken in the

shape of a filament and another electrode is taken in the form of a cylindrical plate. The electrons are attracted by the anode called plate and thus the valve starts conducting.

4. Diode

It is an electronic device or thermionic valve having *two* electrodes. It consists of an evacuated glass bulb containing two electrodes, i.e.,

(i) Plate or anode

(ii) Cathode or filament.

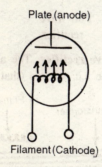

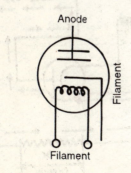

When a current is passed through the filament, it emits a large number of electrons. Such a device is known as directly heated (Fig. 1) and in another device called indirectly heated (Fig. 2) cathode is entirely separate and emits electrons when heated. Diode valves can be used as

(i) rectifier

(ii) detector

Diode valve

| Half-wave rectifier | Full-wave rectifier | Modulator |

Diode as a half-wave rectifier. The above figure shows the circuit diagram. The a.c. voltage (that is to be rectified)

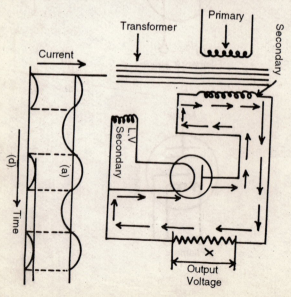

is fed into the primary of a transformer. The positive half of a.c. cycle makes the plate positive and thus the electrons flow to plate (P) and electrons current flows through the plate filament circuit in the direction as shown in figure above. The plate potential becomes during the negative half of the cycle so the electrons are repelled and no current flows.

A voltage appears *across the load x* during positive half of each cycle. Since the voltage appears across the load during half of each cycle, it is called *half-wave rectifier*.

Diode as rectifier. *Rectification* is the process of converting a.c. into d.c. A diode can be used as a rectifier since it permits the current to flow only from cathode to anode (plate) when the plate is at positive potential w.r.t. cathode and no current can flow in the reverse direction.

Diode as full-wave rectifier. This can be achieved with a double made up of two plates along with a filament taken inside the same vacuum glass bulb.

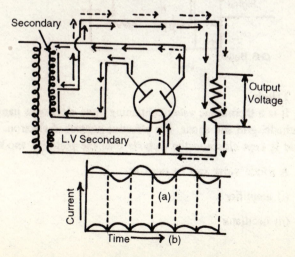

The circuit diagram is shown above. The arrow heads show the direction of current.

During the first half of a.c. The plate P_1 acquires a positive potential and electrons are attracted towards P_1 only so the electronic current flows. During the next half P_2 acquires a positive potential and electrons are attracted towards P_2 only and the electronic current flows as shown in Fig. below:

Thus, the current flows in the same direction through load X in each half of a.c. cycle.

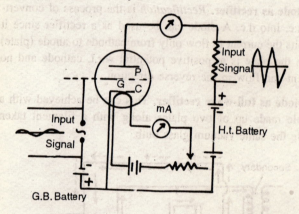

5. Triode

It is a thermionic value containing three electrodes namely cathode, grid and anode. For effective control of electrons the grid is kept closer to the cathode than the plate (or anode).

A triode valve can be used as

(i) amplifier

(ii) oscillator

(iii) detector or modulator.

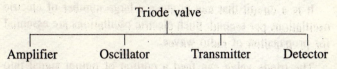

Triode as an Amplifier. It is shown in Figure. The a.c. signal to be amplified is applied across the grid.

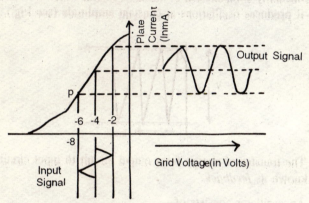

The alternate half cycles of the a.c. signal will swing the grid voltage above and below its bias level and so it causes the plate current to swing in sympathy above and below its static value. In case the amplified a.c. signal is not a correct wave form of input signal then the output signal will be distorted which will blurr the T.V. picture and the radio music. In order to amplify signal without distortion the triode used should be such that its characteristic curve has a large straight portion and in the linear portion of the characteristic, equal changes in grid voltage cause proportionately equal changes in the plate current.

6. Oscillatory Circuit

It is a circuit that can produce a large number of electric oscillations per second. Such electric oscillations are essential for propagation of radio waves.

The triode valve can feed a portion of output signal into input signal in proper phase. The energy loss of tank circuit is constantly compensated at the same rate at which it is lost. So it produces oscillations at constant amplitude (see Fig.).

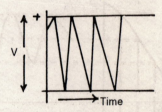

The transfer of energy from output circuit to input circuit is known as *feedback*.

An oscillator consists of—

(i) tank circuit of L — C circuit,

(ii) a correctly phased feedback and amplification,

(iii) a high tension d.c. battery.

A oscillator that produces high frequency oscillations is used as *transmitter* in radio station.

7. Transmitter

To transmit a message the audio frequency oscillations have to be combined with carrier waves. For this purpose the following electric circle is used.

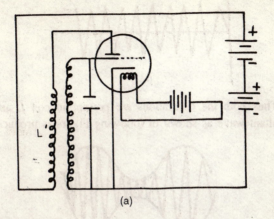

(a)

The sound to be transmitted is produced before microphone M. Microphone is connected to L_3 (a coil) through a battery and a key. This coil (L_3) is so adjusted that it is coupled to another coil L. When we produce sound before microphone the diaphragm vibrates due to which current in L_3 changes continuously. These oscillations are audio frequency oscillations as shown in Figure below :

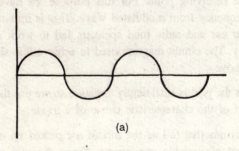

(a)

The triode which acts as an oscillator produces radio frequency as shown in Figure below :

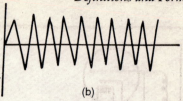

(b)

Thus both the oscillations are present in coil L and the resultant wave as shown in following Figure is produced.

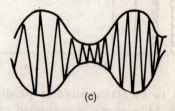

(c)

This wave is called amplitude modulated wave. These oscillations are induced in another coil (L_2) and air transmitted by antenna in all directions.

Triode as detector or rectifier. The speech or music etc., have to be extracted from the modulated waves when they reach the receiving point. For this purpose we have to get audio frequency from modulated wave. This is important because our ear and radio loud speakers fail to work at radio frequency. The circuit diagram used to achieve it is shown in Figure above.

In this the grid is sufficiently negative so we use the lower bend part of the characteristic curve of a triode.

The signals that fall as the areials are picked up with the adjustment of variable condensers. Since the grid is sufficiently negative during the positive half of the cycle of signal voltage so there is a plate current but practically zero plate current during negative half cycle. In this way the lower portion of the cycle is cut and so the current is rectified.

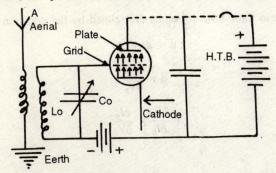

Amplification factor (μ). It is the ratio of the change in plate potential to the change in grid potential needed to produce the same change in plate current.

$$\mu = \frac{\delta V_p}{\delta V_g}$$

$$= \frac{\text{Change in plate potential}}{\text{Change in grid potential}}$$

(keeping plate current constant).

Plate resistance (r_p). It is the ratio of change in plate potential to change in plate current keeping the grid potential constant.

$$r_p = \frac{\delta V_p}{\delta I_p}$$

8. Transconductance or Mutual Conductance (g_m)

It is the ratio of change in plate current to the change in grid voltage producing it, when the plate voltage is held constant.

$$g_m = \frac{\delta I_p}{\delta V_g}, \text{ keeping } V_p \text{ constant.}$$

These three tube constants are related by the equation

$$\mu = g_m \times r_p$$

or
$$\mu = \frac{\delta V_p}{\delta V_g}$$

$$= \frac{\delta V_p}{\delta I_p} \times \frac{\delta I_p}{\delta V_g}.$$

38

X-Rays and Photoelectric Effect

1. X-Rays

When electrons accelerated through a p.d. of V volts strike a target, then the maximum frequency f_{max} of X-rays given off i given by the relation

$$hf_{max} = eV$$

where e is the electronic change.

Production, Properties and Uses

When a beam of highly energetic electrons is allowed to fall on a target with large atomic number, X-rays are produced. The X-rays were discovered by Roentgen in 1895.

Roentgen X-ray production tube suffers from many defects like (i) intensity and quality of X-rays cannot be controlled and (ii) softening device is not perfect. These defects have been removed in modern Coolidge X-ray tube. The cathode and anti-cathode or the target, enclosed inside an evacuated glass tube are connected to high potential drop. Cathode is heated by a battery or a low voltage transformer. X-rays are formed when beam of electrons from the heated cathode are stopped by the anti-cathode or the target. The greater the intensity of electron beam the greater is the intensity of X-rays produced. The quality of X-rays depends upon the voltage applied across the tube. The highest voltage applied is about a million volts, and X-rays of 10^{-10} cm wavelength are produced. The larger the voltage, the harder as the X-rays (smaller wavelength, more penetrating).

X-rays are electromagnetic waves whose wavelengths range between 1 A.U. to 100 A.U. They travel with a velocity of 3×10^8 m/sec in vacuum or air. X-rays can (i) ionize atoms and molecules of matter, (ii) cause fluorescence in some chemical compounds, (iii) affect photographic films, (iv) penetrate through matter (v) show wave-like properties such as interference, diffraction and polarization, (vi) can show particle-like properties in interacting with matter as in photoelectric effect and in Compton effect, and (vii) produce secondary X-rays.

X-rays find many useful and important applications in engineering, medicine, surgery and solid state studies.

3. Photoelectric Effect

It is phenomenon of emission of electrons from a substance under the interaction of electromagnetic radiation with it. The emitted electrons are known as 'photoelectrons'. Certain substances emit photoelectrons when exposed to radiation of suitable frequency only. For example, alkali metals can emit photoelectrons when visible and ultraviolet light is incident on them. X-rays can eject photoelectrons from heavy elements. The important characteristics of the photoelectric effect are :-

Definitions and Formulae in Physics

(i) The number of photoelectrons emitted per second is directly proportional to the intensity of the incident light.

(ii) There is no measurable time lag between the falling of light and the emission of photoelectrons.

(iii) There is a definite frequency of incident radiation, known as the threshold frequency v_0, below which no photoelectrons are emitted, however large the intensity of light may be. Above this frequency even the faintest light can produce photoelectrons proportional to the intensity of light.

(iv) The velocity of photoelectrons ejected is directly proportional to the frequency of the incident light, and is independent of the intensity of light.

The quantum theory of photoelectric effect was given by Einstein in 1905. According to this theory, electromagnetic radiations carry energy in bundles of units hv, where h is Planck's constant and v is the frequency of radiation. these energy packets are known as 'photons'. When a photon of energy hv is incident on a metal surface, its entire energy is imparted to an electron of the metal. A part of this energy is used in liberating the electron from the surface of the metal and the other part is used in imparting a kinetic energy equal to $1/2\ mv^2$ the ejected electron. Hence v is the velocity with which the electron leaves the metal surface, m is the mass of the electron. Hence, the maximum kinetic energy E_k of the electron is given by

$$E_k = \frac{1}{2}\ mv^2 = hv - W$$

where W represents the minimum energy required by the electron to escape from this surface. The minimum energy is known as work function of that surface. The minimum frequency v_0 is related to the work function W as

$$hv_0 = W$$

Hence,

$$1/2 \, mv^2 = hv - hv_0 = h(v - v_0).$$

This equation is known as *Einstein's equation* of the photoelectric effect. For $v = v_0$, kinetic energy is zero. Thus, for frequency $<v_0$, no photoelectrons are emitted. The minimum frequency v_0 below which no photoelectric effect can be observed is known as *threshold frequency*.

Photocell. It is an arrangement used to convert light energy into electrical energy. It is put to many uses, for example, in Television sets, light switches, counting machines, burglar alarm, photometry, etc. These are of two types.

(a) Vacuum type, (b) Gas-filled type.

Laws of Photoelectric Emission

I. The number of photoelectrons ejected is directly proportional to the *intensity of the incident light* and not on the frequency of incident light.

II. The velocity of the escaping electrons is directly proportional to the *frequency of the incident light* and not to the intensity of incident light.

III. Light of frequency greater than the critical frequency or threshold frequency ejects electrons of different velocities.

The *threshold frequency* is the minimum frequency that can eject an electron out of the metal.

Definitions and Formulae in Physics

Einstein's Photoelectric Equation

$$hv = \frac{1}{2} mv^2 = \varphi_0,$$

where φ is a **work** function.

$$\varphi_0 = hv_0$$

(v_0 is the **threshold** frequency)

$$\therefore \quad hv = \frac{1}{2} mv^2 + hv_0$$

or
$$h(v - v_0) = \frac{1}{2} mv^2$$

3. Compton Effect

The **incident photon** carrying an energy hv transfers some of its **energy to the** electron with which it collides, the scattered photon must have a lower energy E', it must therefore, have **a lower frequency** $v\propto = E\propto/h$, thus it has a larger wavelength L'.

The reduction in energy of a photon because of its interaction with free electron is called Compton effect. Part of Photon's energy is transferred to the electrons (called compton or recoil electron) an part is redirected as a photon of reduced energy (compton scatter).

39

Radioactivity

1. Basic Terms

Radioactivity. The rate of disintegration (that is the number of atoms disintegrating per second) is proportional to the number of atoms present. Thus

$$\frac{dN}{dt} = -\lambda N$$

where $\frac{dN}{dt}$ is the rate of disintegration and N is the number of atoms present. λ is a constant called the *disintegration constant*.

Half-life. The half-life of a radioactive substance is defined as the time in which half of any large sample of identical nuclei will undergo disintegration. Half-life is denoted by the symbol T.

If T is the half-life, then the number N of atoms that remain intact after time nT is given by the relation $N/N_0 = (1/2)^n$.

(where N_0 is the initial number of atoms).

If N is the number of atoms that remain after time t, then

$$N/N_0 = (1/2)^{t/T}$$

where N_0 is the initial number of atoms and T is the half-life.

Also the disintegration constant $\lambda = \log_e 2/T$.

The average life of the radioactive atom $= 1/\lambda$.

Nuclear force. The fundamental, natural force which holds the nuclear particles together in the nucleus of an atom.

Nucleon. A general term applied to neutrons and protons, the constituents particles of an atomic nucleus.

Atomic number. A number used to identify elements. It is equal to the number of protons in the nucleus and is represented by the symbol Z.

Mass number. The total number of nucleons in an atomic nucleus. The mass number is the sum of the atomic number and the number of neutrons ($A = Z + N$).

Atomic mass unit. A unit of mass which is equal to one-twelfth the mass of the most abundant form of the carbon atom. Its value is equal to 1.6606×10^{-27} kg.

Mass spectrometer. An instrument used to separate the isotopes of elements based on the differences in mass.

Mass defect. The difference between the rest mass of a nucleus and the sum of the rest masses of its constituent nucleons.

Activity. The rate of disintegration of an unstable isotope. An activity of one curie is equivalent to 3.7×10^{10} disintegrations per second.

Half-life. The time in which one-half the unstable nuclei of a radioactive sample will decay.

Nuclear reaction. A process by which nuclei, radiation, and/or nucleons collide to form different nuclei, radiation, and/or nucleons.

Conservation of charge. The total charge of a system can be neither increases nor decreased in a nuclear reaction.

Conservation of nucleous. The total number of nucleons in a nuclear reaction must remain unchanged.

2. Laws of Radiations

I. The number of atoms that break up at any instant is proportional to the number present at that instant.

Mathematically, $\quad -\dfrac{dN}{dt} \lambda N_0,$

where N_0 = Total number of atoms present
N = Number of atoms present at any resistant of time 't'
λ = disintegration constant or decay constant or radioactive constant.

The above relation can also be written as

$$N = N_0 e^{-\lambda t}.$$

II. The disintegration or radioactive substance is random.

III. The rate of decay of a substance is directly proportional to the number of atoms present at that instant.

Definitions and Formulae in Physics

λ is independent of pressure, temperature, humidity, etc.

Radioactive constant 'λ. It is given by

$$\lambda = \frac{1}{t}$$

so it is defined as number of atoms decaying per second.

From the equation

$$N = N_0 e^{-\lambda t}$$

we get

$$\frac{N}{N_0} = e^{-\lambda t}.$$

Half-life or half-life period ($T_{1/2}$). It is the time required for the disappearance of one half of the amount of the radioactive substance originally present.

Mathematically, $\quad T_{1/2} = \dfrac{0.693}{\lambda}.$

Alpha particle. The nucleus of a helium atom, which consists of two neutrons and two protons bound together by nuclear forces.

Beta particle. A beta minus particle is simply an electron. The beta plus particle has a mass equivalent to the electron, but its charge is equal and opposite to that of an electron.

40

Nucleus and Nuclear Energy

1. Nuclear Physics

The nucleus of an atom consists of protons and neutrons only. The mass of a proton is equal to the mass of the nucleus of an atom of ordinary hydrogen and it has a positive charge equal to that of an electron. Neutron has the same mass as proton but it has no charge.

Mass number. The mass number of a nucleus is the number of protons and neutrons present in it.

Atomic number. The atomic number of an element is the number of protons contained in its nucleus.

The usual method of representing a nucleus is illustrated below for oxygen :

$$_8O^{10}.$$

Here the mass number is indicated by the superscript (on the right top) and the atomic number is given as subscript (on the left bottom). Isotopes of an element have the same atomic number but different mass numbers. The isotopes of an element have the same number of protons in their nuclei but they have different number of neutrons.

The binding energy. The binding energy of a nucleus is the energy equivalent of the difference between its mass and the sum of masses of its individual components (protons and neutrons).

Equations for nuclear reaction. The following rules for balancing equations for nuclear reactions should be remembered :

(i) In a balanced equation the sum of the subscript (atomic-numbers) must be the same on the two sides of equation.

(ii) In a balanced equation the sum of the subscripts (mass numbers) must be the same on the two sides.

For example, $_7N^{14} + {_1H^1} \to {_6C^{11}} + {_2He^4}$

2. Nuclear Fission

It is the splitting of a heavy nucleus into two approximately equal parts due to capture of a neutron by the nucleus. In fission a lot of energy is set free. A typical example is the fission of $_{92}U^{235}$:

$$_{92}U^{235} + {_0n^1} \to {_{92}U^{236}} \to {_{54}Xe^{140}} + {_{38}Sr^{94}} + {_0n^1} + y + 200 \text{ Mev.}$$

3. Chain Reaction

Because each fission liberates on the average two neutrons while only one is required to initiate fission, a rapidly multiplying sequence of fission can occur in a lumb of suitable material thus setting up a chain reaction which sets free an immense amount of energy within a short time.

4. Nuclear Fusion

The binding together of light nuclei to form a heavier nucleus is called nuclear fusion. Here too a lot of energy is released. This energy is called **thermonuclear energy**. Fusion reaction can take place only at very high temperatures.

Example of fusion reaction

$$_1H^2 + {_1H^2} \rightarrow {_1H^3} + {_1H^1} + 4.0 \text{MeV}.$$

Here two deuterons ($_1H^2$) fuse together to form a triton ($_1H^3$) and a proton ($_1H^1$).

5. Electron Volt (eV)

It is the amount of work done when one electron moves through a potential difference of 1 volt.

$$1 \, eV = 1.6 \times 10^{-12} \text{ ergs}$$

$$1 \, MeV = 1.6 \times 10^{-6} \text{ ergs}.$$

Also remember

$$E = mc^2$$

$$1 \text{ a.m.u.} = 931 \text{ MeV}$$

$$1 \text{ a.m.u.} = 1.66 \times 10^{-27} \text{ kg}.$$

6. Nuclear Reactor

It is a device in which enormous nuclear energy in produced by controlled chain reaction. These are also called atomic piles. There are two types of nuclear reactors.

(i) *Homogenous reactors.* In these heavy water is used as moderators.

(ii) *Heterogeneous reactors.* In these graphite rods are used as moderators.

Breeder reactor. It produces power due to fission by fast neutrons. It simultaneously regenerates more fissionable material than it consumes.

Moderator. Any substance that slows down the fast moving neutrons in a nuclear reaction is called moderator, e.g., heavy water, graphite, beryllium oxide, ordinary water, etc.

Proton. It has unit positive charge and a mass equal to 1.6725×10^{-27} kg.

Neutron. It has no charge and a mass equal to 1.675×10^{-27} kg.

Electron. It has a unit negative charge and negligible mass.

Positron. It was discovered by Anderson in 1932.

It is called antiparticle because it can annihilate an electron.

$${}^0e_{+1} + {}^0e_{-1} = 2\ h\nu \text{ (pair annihilatron)}$$
$$\text{positron} \quad \text{electron}$$

7. Nucleus

The positively charged core of an atom, consisting of one or more protons and except in case of hydrogen, one or more neutrons. Whole of the mass of atom is concentrated in the nucleus. It occupies only a tiny fraction of the volume of the atom.

8. Isotopes

The atoms of an element having the same atomic number and identical chemical properties but different atomic weights are known as isotopes of that element :

Hydrogen : $^1H_1$, $^2H_1$, $^3H_1$

Oxygen : $^{16}O_8$, $^{17}O_8$, $^{18}O_8$.

9. Primary Cosmic Rays

Cosmic rays are entering our atmosphere constantly from all directions of outer space. They consist of atomic nuclei obtained from atoms which have lost their electrons. They consist of 92% protons, 7% α-particles (Helium ions He^{2+}) and rest bare nuclei of light elements. Their energies range from 10^9 to 18^{18} eV.

10. Secondary Cosmic Rays

They are mostly μ mesons and positive and negative electrons. These are produced when primary cosmic rays collide with air molecules.

41

Universe

1. Astronomy

Astronomy is that branch of science which deals with the study of heavenly bodies, viz., the earth, the moon, other planets, their satellites, the sun and other stars. Astronomy is studied with the help of observations rather than any other experimental ways as the heavenly bodies are not in our control unlike the experimental objects in case of traditional set ups.

2. Constituents of the Universe

It is estimated that there are several billions of billions 'Milky Ways' in the universe each of which is a cluster of millions of stars, one such star being our sun. These milky ways, as they are popularly called, are termed as different galaxies in scientific terminology. Next, like any other stars, our sun has some planets moving about it. These are : Mercury, Venus, Earth, Mars, Jupiter, Satron, Uranus, Neptune and Pluto in the increasing order of distance from the sun. Further, some of these planets have some moons of their own revolving about them, e.g., our earth has one moon, satron has ten moons, mars has two moons, jupiter has fourteen moons, uranus has five moons and neptune has two moons as per the latest records of discoveries.

3. Instruments for Astronomical Studies

There are various instruments used for the study of the astronomy, telescopes are the most important. These are of two types, viz., the refracting ones and the reflecting ones. Reflecting telescopes have several advantages over the refracting types like these do not have associated chromatic abberation, large aperture mirrors are more easily available than lenses etc.

Observations of radiosignals coming from far off heavenly bodies is done with the help of radio-telescopes. Radio telescopes are capable of catching radio signals which, being of much higher frequency than light, can travel through much larger distances on account of their higher energy content, and lower attenuation. Additionally, as radio signals do not get disturbed by the daylight, one can work with radio signals during the day too, unlike the conventional telescopes which are used at night.

4. Kepler's Laws of Planetory Motion

The three laws of planetory motion given by Kepler are as follows :

(i) All planets in our solar system move around the sun, which always remains stationary, in elliptical orbits such that sun lies at one of the focii of the orbits.

(ii) the aereal velocity which is the area traced out per unit time by a planet is constant or

$$\frac{\Delta A}{dt} = \text{Constt}$$

Definitions and Formulae in Physics

This demands that when the planet is near the sun, it should move fast when it is away from the sun, it should move slow.

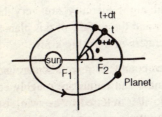

(iii) The square of the time period of the planet is directly proportional to the cube of the semi-major axis of its trajectory.

5. Our Solar System

Our solar system revolves around the sun. For all practical purposes sun can be treated as a stationary star. The mass of the sun is estimated to be 1.98×10^{30} kg and radius 6.956×10^8 m. Further, the mean temperature of the sun is around $6000°K$ and thermonuclear fusion reactions are taking place in the sun. The energy released in these reactions is being emitted by the sun continuously in all directions. Sunlight is the chief source of energy on all he planets revolving about the sun.

Besides the sun, our solar system has nine planets with some of these having different number of moons of their own revolving around them.

6. Some Important Planets

Some of the important planets along with their important characteristics are as follows :

Mercury : Mercury completes one revolution around the earth in about 88 days or so. Experimental studies of this planet reveal that there is no atmosphere on the mercury and hence there is no life in it. Also, temperature variations during day and night mercury are very large say a few hundred degrees and the surface of this planet is very uneven with irregular craters of very big sizes.

Venus : Venus completes its rotation around the sun in 225 days. Its size, mass and radius are roughly equal to those of the earth and so it is referred to as twin sister of the earth also. The atmospheric pressure on the moon is about 100 times that on the earth and the atmosphere mainly consists of carbon dioxide which is about 95%. The water content is about 1% and the mean temperature is about 480°C.

Mars : Mars completes one revolution about the sun in 690 days or so. It's atmosphere is not very dense and it is also composed of carbon dioxide mainly with traces of water. Initially it was believed to have some life on it. However, results of the latest studies reveal that there is no life at all on mars.

Other Planets : Other planets being very far off from the sun are so cold that there is no question of any life on them. Also, experiments reveal that most of them have atmosphere consisting of poisonous gases like ammonia and methane etc.

7. Moon

Moon is natural satellite of the earth. Besides earth, some other planets have different numbers of moons of their own. Moon revolves around the earth in one full day, i.e., twenty four hours. There is no atmosphere on the moon on account of low gravity. The temperature of the moon varies from

110°C to —150°C or so. the surface of the moon is irregular with lots of craters distributed all over. There is no life on moon.

Of the other planets Mars has two, Jupiter has fourteen, Satron has ten, Uranus has five and Neptune has two moons respectively.

8. Asteroids

It has been observed that about 1600 heavenly bodies of sizes of the order of 10^{-5} to 10^{-7} times that of our earth are available between Mars and Jupiter going round the sun like other planets. On account of their small sizes these cannot be termed as planets. They are considered to be parts of Mars and Jupiter breaken loose out of the parent planets. They are quite similar in their composition to the moon. These are called Asteroids.

9. Comets

Comets are dense balls of ammonia, methane and other gases, moving in parabolic paths of highly elongated shapes. Some of them appear periodically close to our earth. However, the period of appearance may vary from about 20 years to a few hundred years when a comet approaches the sun, on account of sun's heat, the nucleus of the comet gets detached somewhat and it takes a typical shape with a long tail of gases shinning brightly over a length of 10^3 km or so.

10. Meteors

Small pieces of comets keep on falling out in the space on and off. They look like fire balls when approaching the sun in its bright light. On their way, most of them burn off in the space itself whereas some others fall on different planets. These prices of comets are called meteors. It is assumed that some craters on different planets must have been caused by the impacts of some meteors falling on them at different times.

11. Some Elementary Measurements in Solar System

(i) **Distances** : Measurement of distances are done either by triangulation method or with radar echo method. In the triangulation method, angles of elevation of the distant object are measured at two positions separated by a known distance later using simple trigonometrical relations, if both the dis-

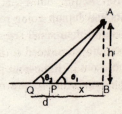

tance of the object as well its height from earth's surface, i.e., h and x can be calculated. In radar echo experiments, radio waves of extremely small wavelength are sent upto the object. These get reflected and the reflected beam is received back. Knowing the time lag between the time of sending and receiving back of these waves, with the helps of velocity of electromagnetic waves, one can calculate the distance of the object from the earth.

(ii) **Sizes** : Sizes of objects can be similarly calculated by finding out the angle subtended by the objects at the earth using sextant. It the angle subtended is 'θ' and the distance between the object and earth is 'r' the diameter of the object would be

$$d \simeq r.\theta.$$

Here θ is to be strictly taken in radians.

(iii) **Period of rotation** : Period of rotation of different planets can be calculated through Kepler's third law according to which

$$T_1^2/T_2^2 = a_1^3/a_2^3$$

where T_1 and T_2 are periods of the planets moving in elliptical of semi major axes a_1 and a_2. Thus, knowing the semi major axes of the orbits of two planets and knowing the period of one of them, period of the other can be deduced from the above relation.

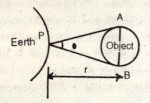

(iv) **Mass of sun** : Mass of the sun, M, is given by relation, $M = \dfrac{4\pi^2 r^3}{GT^2}$ where r is the radius of the earth, T its time and G its gravitational path.

(v) **Surface temperature.** Surface temperature of the sun can be estimated by measurement of what is called as the Solar constant. Solar constant denoted by S can be defined as the amount of energy received from the sun per unit area of the earth's surface per unit time. Experimentally, it comes out to be 1388 Joules/sec/m³. Surface temperature is given by the expression

$T = [(R/r)^2 \times S/\sigma]^{1/4}$ and it comes out to be 5730°K.

(vi) **Atmosphere.** The study of the atmosphere of a planet is done with the help of determination of a factor called Albedo. Albedo is the ratio of the amount of light energy reflected or rather radiated by a planet to the amount of energy incident upon it. Denser is the atmosphere available on the planet, larger is the albedo ratio expected as the atmosphere must reflect the light energy incident on the planet.

12. Study of the Sun

Sun is the centre of our solar system. It is the nearest of all stars and so it appears brightest of all starts to us. Light coming from the sun takes about eight minutes to reach us. Mass of the sun is about 33000 times that of the earth whereas it's radius is about 108 times that of the earth. The whole of the sun is expected to be composed of gases of which about 70% is hydrogen and 28% is helium. Temperature of the inner most part is around 14×10^6 degrees Kelvin and that of the surface is about 5700 degrees Kelvin.

13. Stars

Stars are stationary bodies of much larger mass than any planet of the given solar system, around which the entire solar system revolves. As there are about 10^{11} galaxies and every galaxy has in turn about 10^{11} solar systems, obviously 10^{22} stars are expected to constitute the universe. The farther is the star from us, fainter it will appear to us.

14. Life of a Star

It is expected that stars keep on getting formed or born and also keep on getting disintegrated or dying like any human beings on the earth. Thus, some of the stars are young, others youth and some others old ones also. Initially when wandering clouds of gases (mainly hydrogen) collect together to the tune of about thousands times of a solar system, on account of gravitational pull, they come close to each other and form a compact system and thus a star is born. Initially, the temperature of the star is as low as $-173°C$ or so. However, the molecules of the gases start making inter collissions which results in increase in temperature which with the passage of time may rise upto 10^7 degrees or so. Such a high temperature leads to the start of fusion reactions in the central part of the star. These reactions result in the production of hormons about of energy which the star starts radiating. The star is now fully grown up one. Our own sun is in this stage. The star continues to be in this stage for a few hundred to a few thousand million years after which all the hydrogen content is expected to finish through fusion reactions. Thereafter, its decay starts. It commences with the cooling of it's outer surface and it's expansion thereof. This is accompanied with greater brightness and in this stage it is called to be *Red Giant*. After the red giant stage, there are two ways of decay of a star. If it's mass is smaller than that of our sun, it's central part contracts further to make what is called a *White Dwarf*. The central part of the white dwarf is dense matter with electron cloud surrounding it. With passage of time, the white dwarf is expected to cool down and convert itself into a small non-active opaque to all radiations very dense body called a *Black Dwarf*. On the other hand, if the mass of the star is more than that of our sun, it explodes after red giant stage creating what is called a *Supernova*.

15. Galaxies

Millions of billions of star together form a galaxy. Besides all the stars, planets, moons, a galaxy also has it shinning clouds at gases called *Nabulae*. A galaxy is also referred to as *Milky Way* on account of it's shining character.

16. Theories of Cosmology

Cosmology is the study of age, mass origin and evolution of universe. Studies in cosmology reveal that universe in ever expanding in the form of continuous drifting of all galaxies from one another. According to Hubble's law, the speed of drift of a galaxy from another one is directly proportional to the distance between them i.e., $k \times v$, where v is the velocity of drift and r is the separation between the galaxies. There prevalent theories of formation and evolution are

(i) **Big bang theory** : As per this theory, entire universe used to be a big sphere of solid mass millions of years ago when a big explosion took place. The fragments of this big mass went flying in different directions.

(ii) **Pulsating theory** : According to this theory, the universe is alternately expanding and contracting. Presently, we are passing through the period of expansion and after sometime which may be a few million years, further deift will stop and contraction will start taking place which may keep on for a few million years beyond which expansion may again start occurring and so on.

(iii) **Steady state theory** : As per this theory, the overall size and mass of the universe is neither changing nor reshaping. Formation of new stars and galaxies and dying out of ones are minor changes which, of course, keep on taking place.

42
Solids

1. Solids

A solid is characterised by the fact that intermolecular forces are the strongest in it and so it retains its shape and volume normally.

Solids are classified as crystalline and non-crystalline or amorphous. The distinguishing features of these two forms are as follows.

(i) **Crystalline solids.** In crystalline solids, all atoms are arranged in a regular way in a repeated manner in what is called a lattice. They have fixed and well defined i.e., sharp melting points. Crystalline solids sometimes exhibit an isotrophy in the matter of elastic properties, optical activity, electrical conductivity etc., which means they can have different elastic moduli, electrical resistivity etc. in different directions. Typical examples of crystalline solids are diamond, sodium chloride, lead etc.

(ii) **Non-crystalline or amorphous solids.** These may be in powdered form or otherwise. Atoms are not arranged in any fixed or particular order in them. There is no question of any isotropic behaviour, therefore. Some examples of amorphous solids are graphite, glass, rubber etc.

2. The Crystal lattice.

Crystal lattice is a geometrical representation of the positions of various atoms of the crystal. It may be both two dimensional or three dimensional. As in a lattice, the atoms are regularly placed, assuming one of the atoms to be the reference origin, it is possible to express positions of other atoms in the form, $\vec{r} = n_1\vec{a} + n_2\vec{b} + n_3\vec{c}$ where n_1, n_2 and n_3 are integers and $\vec{a}$, $\vec{b}$ and $\vec{c}$ are characteristic vectors of the given lattice called translational vectors in three dimensions. The adjoining figure shows a two dimensional lattice and ore can see that for the atom as A, $\vec{r} = 2\vec{a} + 3\vec{b}$.

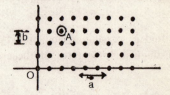

3. Unit Cell

A unit cell is the smallest part of the lattice which is complete in itself and whose repetition on all sides and in all planes creates the lattice in full.

4. Coordination Number

Coordination number of a unit cell is the total number of nearest neighbouring atoms to a particular atom of the lattice.

Definitions and Formulae in Physics

5. Atomic Radius

It is defined as the radius of circle with two nearest neighbouring atoms in the unit cell forming ends of the diameter of that circle. Thus, it is half the distance between the nearest neighbouring atoms.

6. Atomic Packing Factor

It is defined as the ratio of the volume of the sphere of radius equal to the atomic radius to the volume of the unit cell. It gives an idea of the closeness of atoms in the unit cell.

7. Cubic Crystals

A cubic crystal has it's unit cell in the form of a cube. There are three basic forms of a cubic crystal viz., simple cubic, body centered cubic and face-centered cubic. In a simple cubic, there is one atom on each of the six corners of the cube. In a body centred cubic, besides the above, there is an atom first at the centre of the cube while in a face centered cubic, besides those atoms at the corner, there is one atom in the centre of each of the six faces.

Simple Cubic   Body Centred Cubic   Face Centred Cubic

8. Characteristics Parameters of Cubic Crystals

There are five characteristic parameters of the cubic crystals which in the case of the above three compare as follows :

It is to be noted that every corner atom is shared by 6 unit cells and so it contributes (1/6th) of atom to one unit cell.

Similarly, every atom in the middle of a face is shared by two faces and so it contributes (1/2) of an atom per unit cell. While one in the middle of the cube wholly belongs to that unit cell itself.

S. No.	Crystal type	Volume of unit cell	No. of atom per unit cell	Coordination no.	Atomic radius	Packing factor
1.	Simple cubic	a^3	1	6	$\dfrac{a}{2}$	$\pi/6$
2.	Body centered cubic	a^3	2	8	$\dfrac{\sqrt{3}}{4}a$	$\dfrac{\sqrt{3}}{6}\pi$
3.	Face centered cubic	a^3	4	12	$\dfrac{a}{2\sqrt{2}}$	$\dfrac{\sqrt{2}}{6}\pi$

9. Packings in Crystals

There can be many types of closed packings of crystals of which the hexagonal closed packing is the simplest. In the two dimensional plane, it looks like in Figure (a), while in three dimensions, it looks like in Figure (b).

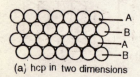

(a) hcp in two dimensions

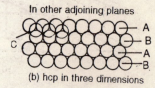

(b) hcp in three dimensions

10. Bonding in Solids

There are four types of chemical bondings usually observed in solids viz., the ionic bonding, covalent bonding, metallic bonding and Van-der-Waal's bonding. Some important features of these bonds are follows :

(i) **Ionic bonds.** In case of ionic bonds, there is a transfer of electrons from one type of atoms to other type. This means that the electronic configuration of a outermost orbits is changed and some atoms lose electrons whereas others gain. All alkali halides e.g., NaCl, KBr etc. are typical example of these type of solids. The atoms on losing or gaining electrons become positive and negative ions. Thus, NaCl consists of Na^+ and Cl^- ions. It is the attraction between positive and negative charges of the ions which is responsible for this type of strong bonding. Ionic bonding is very strong and such solids are very hard, brittle, poor conductors of electricity in solid state and have high heat of vaporisation and high melting points.

In relation to ionic bonding, the term cohesive energy of a crystal is defined as the energy required to break up the crystal into respective ions.

(ii) **Covalent bonds.** Covalent bonds occur due to the sharing of electrons by more than one atoms. Thus, unlike the ionic bonds, there is no actual transfer of electrons from atom to atom. Silicon, germanium, diamond, etc., are typical examples of covalent bonded crystals. Covalent bonds are also quite strong and crystals formed by such bonding are again very have and brittle and hard high melting point but they are poor conductors of electricity.

(iii) **Metallice bonds.** Metals are characterised by the presence of free electrons in them. Actually most of the metals are high atomic numbered elements. The electrons in the last orbits of there atoms are, therefore, very loosely bound to the nucleii and sometimes their mutual colissions result in transfer of energy which is food enough for these electrons to come out of them orbits, thus giving rise to availability of free electrons. These free electrons are free in the sense that they are not bound to respective atoms and can freely travel within the solid in all directions. However, they cannot come out of the surface of the solid as they are bound with what is called as surface barrier. Metals which mostly are characterised by the presence of these bonds are good conductors of heat and electricity both, are isotropic in behaviour and have shining appearance.

(iv) **Van der Waal's bonds.** These are weak bonds on account of attraction amongst atoms and molecules which do not have a uniform distribution of charge on them. Thus, if the centre of mass is different from the seats of positive and negative charges, this type of bondings will take place. In other words, attractive forces amongst dipoles are responsible for Van-der-Waal's bonds. Most of the gases and soft solids have this type of bondings. Some characteristics of there bonds are that they are very weak, there are no free electrons and so these are poor conductors.